Eugenics, Literature and Culture in Post-war Britain

Routledge Interdisciplinary Perspectives on Literature

1 **Environmental Criticism for the Twenty-First Century**
Edited by Stephanie LeMenager, Teresa Shewry, and Ken Hiltner

2 **Theoretical Perspectives on Human Rights and Literature**
Elizabeth Swanson Goldberg and Alexandra Schultheis Moore

3 **Resistance to Science in Contemporary American Poetry**
Bryan Walpert

4 **Magic, Science, and Empire in Postcolonial Literature**
The Alchemical Literary Imagination
Kathleen J. Renk

5 **The Black Female Body in American Literature and Art**
Performing Identity
Caroline A. Brown

6 **Narratives of Migration and Displacement in Dominican Literature**
Danny Méndez

7 **The Cinema and the Origins of Literary Modernism**
Andrew Shail

8 **The Gothic in Contemporary Literature and Popular Culture**
Pop Goth
Edited by Justin D. Edwards and Agnieszka Soltysik Monnet

9 **Wallace Stevens and Pre-Socratic Philosophy**
Metaphysics and the Play of Violence
Daniel Tompsett

10 **Modern Orthodoxies**
Judaic Imaginative Journeys of the Twentieth Century
Lisa Mulman

11 **Eugenics, Literature and Culture in Post-war Britain**
Clare Hanson

Eugenics, Literature and Culture in Post-war Britain

Clare Hanson

Routledge
Taylor & Francis Group
NEW YORK LONDON

First published 2013
by Routledge
711 Third Avenue, New York, NY 10017

Simultaneously published in the UK
by Routledge
2 Park Square, Milton Park, Abingdon, Oxon OX14 4RN

Routledge is an imprint of the Taylor & Francis Group, an informa business

Library of Congress Cataloging-in-Publication Data
Hanson, Clare.
Eugenics, literature, and culture in post-war Britain / by Clare Hanson.
p. cm. — (Routledge interdisciplinary perspectives on literature ; 11)
Includes bibliographical references and index.
1. English literature—20th century—History and criticism. 2. Literature and society—Great Britain—History—20th century. 3. Eugenics in literature. 4. Eugenics—Great Britain—History—20th century. 5. Biopolitics—Great Britain—History—20th century. I. Title.
PR478.S57H36 2012
820.9'00914—dc23
2012009899

ISBN13: 978-0-415-80698-5 (hbk)
ISBN13: 978-0-203-09791-5 (ebk)

Typeset in Sabon
by IBT Global.

Contents

Figures

All images courtesy of the Hulton Getty Archive except the cartoon by David Low, which is courtesy of London Evening Standard.

Acknowledgements

I am grateful to the Arts and Humanities Research Council and the University of Southampton for the research leave that enabled me to complete this book. Thanks are also due to the Wellcome Trust for a research award that supported the initial archival research for this project. I am grateful to my colleagues at Southampton for many conversations about eugenics; in particular I would like to thank Stephen Morton for suggesting key references, and David Glover and Peter Middleton who generously gave their time to read and comment on draft material. I have also benefited from the expertise of Mary Eagleton in connection with the gendered aspects of post-war culture. Some of the material in the book was first presented when I was a Seelye Visiting Fellow at the University of Auckland and at conferences at Leeds Metropolitan University and San Diego State University, and I am grateful to colleagues who attended these events for their comments. Finally I would like to thank my family for their encouragement and support.

Acknowledgements

Introduction

In February 1943, the guest speaker at the Eugenics Society's annual Galton Lecture took up the question of what he called 'the main dysgenic factor in our society today', which was that the middle- and upper-classes were being outbred by the genetically inferior working-classes.[1] There is nothing surprising in this—the topic of differential breeding was a staple for eugenicists—except that the speaker was William Beveridge, the author of the Beveridge Report, which laid the foundations for post-war reconstruction and the Welfare State. Beveridge had arrived hotfoot from the first House of Commons debate on his Report to address a large audience of the Eugenics Society (of which he was a life-long supporter) and to argue that the proposals contained in it could, in the long term, have positive eugenic effects. We are not used to making connections between eugenics and the project of post-war reconstruction, and it is generally assumed that eugenic thought petered out in Britain after the Second World War; this view has been endorsed by some historians of eugenics, including Pauline Mazumdar, who has argued that the eugenic project disintegrated under the pressure of 'changed social attitudes'.[2] As I shall show, this was far from the case. Eugenic thought was a key point of reference for politicians and policy makers in the 1930s, and it remained an important resource both during and long after the war, as policy makers grappled with ways of tackling the 'five giants' Beveridge identified as obstacles to social progress: 'Want, Disease, Ignorance, Squalor and Idleness'. The input from eugenics was facilitated by the fact that several of the chief architects of post-war reconstruction, including Beveridge, Richard Titmuss and C. P. Blacker, were closely involved with the eugenics movement.

This study explores the impact of eugenic thought between 1945 and 1979, a period in which successive governments supported a form of 'welfare capitalism' aimed at reducing inequality through state intervention. To posit a link between eugenics and post-war reconstruction seems counter-intuitive in the context of a progressive post-war settlement, which was being worked out in the shadow of Nazi atrocities. However, the connection becomes more intelligible if we locate it in the context of the longer term history of eugenics, a word first coined by Francis Galton when he defined it in this way:

> Eugenics is the science of improving stock, which is by no means confined to judicious mating, but which, especially in the case of man, takes cognisance of all influences that tend in however remote a degree to give the more suitable races or strains of blood a better chance of prevailing speedily over the less suitable than they otherwise would have had.[3]

Galton's ideas are firmly rooted in evolutionary theory and his cousin Charles Darwin's account of the origin of species; indeed he presents eugenics as a helpful adjunct to natural selection, arguing that by applying the techniques of animal breeding to human beings, modern man can take control of evolution, speeding up the improvement of the human race. As he puts it, '[w]hat Nature does blindly, slowly, and ruthlessly, man may do providently, quickly, and kindly'.[4] However, Galton's interest in heredity is also associated with marked fears about the ability of 'the race' (by which he usually means the Anglo-Saxon race) to withstand the challenges it will face in the future. In this respect, as Soloway has argued, eugenics was from its inception intimately bound up with concerns about the increasing complexity of modern civilisation and the pressures of industrialisation, (post)colonialism and mass democracy: in short, modernity.[5] From Galton's perspective, improving the quality of the population ('quality' being defined in terms of 'ability' and 'energy') was the obvious way to meet these challenges.

In his 1865 essay on 'Hereditary Talent and Character' and in *Hereditary Genius: An Inquiry into its Laws and Consequences*, published four years later, Galton defines the population as a national resource that is currently under threat because the proportion of 'gifted' individuals within it is in decline. In order to reverse this trend, he proposes a wholesale reform of what he describes as 'the haphazard customs of marriage and procreation' and the introduction of a system in which the talented will marry each other to create a 'caste' of the naturally gifted—although he warns that this caste will need to be constantly replenished in order to counteract a tendency to 'revert to meaner ancestral types'.[6] What he envisages, then, is a social hierarchy based on eugenic qualities rather than inherited wealth, in which 'ability' in conjunction with 'energy' will be recognised and rewarded. This is a theme to which he returns throughout his career, arguing in 1909, for example, that '[a]n enthusiasm to improve the race would . . . express itself by granting diplomas to a select class of young men and women, by encouraging their intermarriages, by hastening the time of marriage of women of that high class, and by provision for rearing children healthily'.[7] At first sight these may seem implausibly crude suggestions, yet their influence is clearly discernible in the concepts and policies associated with the post-war meritocracy, which is discussed in the first chapter of this book. The extraordinarily sustained influence of eugenic thought can be explained in part by the fact that the movement rapidly became entwined with the aims of an upwardly mobile middle-class.

Galton's argument in *Hereditary Genius* is that the aristocracy has become biologically weak (surviving only because it is protected from the pressures of natural selection by privilege), while the lowest classes of society are inherently lacking in ability. The reservoir of 'talent' on which the future of the race depends lies in the professional middle-class, and 'gifted' individuals from this class must be enabled to fulfil their potential—an analysis likely to appeal to a section of society that also had the time and resources to support the eugenic cause.

BIOPOLITICS AND EUGENICS

Evolutionary theory is a key point of reference in Michel Foucault's influential analysis of biopolitics, which itself forms part of the wider context of this study. Foucault opens his account of biopolitics with the contention that in the late eighteenth-century, the 'disciplinary' techniques discussed in his *Discipline and Power* (1975), which were associated with the rise of the medical profession and directed at the individual body, were supplemented by a new 'technology of power', which he calls 'regularization'. The object of this power was not individual man, but man at the level of the population, 'affected by overall processes characteristic of birth, death, production, illness and so on'.[8] New techniques were introduced in order to map this object of knowledge/power, including demography and statistics, the long-term aim being to stabilise the population and maintain its equilibrium. Situating these techniques in a broader context in which the old sovereign right 'to take life or let live' has been supplemented by the power 'to make live and let die', Foucault argues for a complex series of transformations whereby what is at stake in power (which may operate outside the state as well as within it, through philanthropy or social reform) has become the security of life itself. This aim of biopolitics, then, is the protection and preservation of what Foucault calls 'the multiple body' of the population:

> What we are dealing with in this new technology of power is not exactly society (or at least not the social body, as defined by the jurists), nor is it the individual-as-body. It is a new body, a multiple body, a body with so many heads that, while they might not be infinite in number, cannot necessarily be counted. Biopolitics deals with the population, with the population as political problem, as a problem that is at once scientific and political, as a biological problem and as power's problem. (245)

Through the image of the multiple body, Foucault echoes and inverts the image on the frontispiece of Hobbes's *Leviathan* discussed in lecture 5 of *Society Must Be Defended*, in which the gigantic head of a king rises out of a torso composed of three hundred tiny bodies. In this version, the head of the king has been replaced by a proliferation of hydra-like heads that threaten to escape classification and control. This is not, however, an image of liberatory potential; in invoking Hobbes at the very point when he defines the population as the object of biopolitics, Foucault underscores the potentially coercive/eugenic aspect of this 'new' form of power. The allusion to Hobbes also sets the tone for his discussion of biological racism, a theme he develops in dialogue with the concept of the 'war of every man against every man', which was for Hobbes more or less identical with man's natural state; for Foucault, the war of man against man pervades every aspect of the *social* body.

Biological racism, Foucault argues, arises out of a fundamental paradox: how can a biopolitical state that is committed to the preservation of life permit or license the right to kill? How can power kill if its basic function is to improve life, to lengthen it, to avoid risks and compensate for its failings? The answer lies in the articulation of a new form of racism that inserts a break into the realm of life, between 'what must live and what must die' (254), dividing the biological continuum into species and subspecies. But this is a specifically modern racism, distinct from earlier struggles in society between rulers and ruled, which were cast in racial terms.[9] Biological racism is not about a struggle for position within society, but it is bound up with the very foundation and legitimation of a modern state that is 'obliged to use race, the elimination of races and the purification of the race, to exercise its sovereign power' (258). According to Foucault, such racism operates on two levels. Firstly, it allows war to be conceptualised not in geopolitical terms but as the destruction of an enemy race, of a biological threat posed by one set of people to another. Secondly, it legitimates the elimination of internal as well as external threats to the population, according to the following logic:

> 'The more inferior species die out, the more abnormal individuals are eliminated, the fewer degenerates there will be in the species as a whole, and the more I—as species rather than individual—can live, the stronger I will be, the more vigorous I will be.' (255)

Foucault puts this passage in quotation marks to signal that this is an expression of inner consciousness, an evocation of the logic of identification with the body politic that is one of the hallmarks of eugenic thought. He thus highlights the imbrication of eugenics and biological racism and draws attention to the way in which eugenic ideas could be—and were—mobilised to legitimate the biopolitical techniques of regularization involved in 'making live' or 'letting die'. Eugenics is also implicated in his analysis of Nazism as the 'paroxysmal' development of state racism (259). As he suggests, in the Nazi state, the 'control of the biological' was unremitting, with regulation of reproduction and control over illness and accidents; this was a regime intent on 'controlling the random element inherent in biological processes'. Yet it was also a society that unleashed murderous power (the old 'right to kill') in the service of biological racism. Foucault writes:

> The Nazi State makes the field of the life it manages, protects, guarantees, and cultivates in biological terms absolutely coextensive with the sovereign right to kill anyone, meaning not only other people, but also its own people. There was, in Nazism, a coincidence between a generalized biopower and a dictatorship that was at once absolute and retransmitted throughout the entire social body by this fantastic extension of the right to kill and of exposure to death. (260)

He here draws attention to the self-cancelling logic of Nazism, in which any given individual could at a stroke move from being the agent to the object of the regenerative programme. For Foucault this ambivalent play between the poles of life and death can be found in almost all modern states, including socialist states. He goes on to coin the term 'social-racism' as a shorthand for the linkage of socialism and biopower and concludes that 'racism—not a truly ethnic racism, but racism of the evolutionist kind, biological racism—is fully operational in the way socialist States (of the Soviet Union type) deal with the mentally ill, criminals, political adversaries, and so on' (261–62).

GOVERNMENTALITY

Foucault returns to the connections between socialism and biological racism in the lectures published as *Security, Territory, Population* (2007) and *The Birth of Biopolitics* (2008) in which he develops his influential account of governmentality. By this he means governmental practice in a post-Enlightenment context in which government is no longer answerable to any external power. In the absence of legitimation by divine or sovereign right, how does government devise its own reason? What are the techniques that process or shape 'reality' in such a way as to make it governable? Foucault identifies a range of mechanisms that operate in three intersecting realms. The first is that of government rationale, that is, the more or less self-conscious reflection of government on itself that draws on a range of expertise in order to conceptualise human beings in such a way that they can be governed. The second is that of practical intervention, the complex array of policies, programmes and mechanisms that can be mobilised to modify human conduct. Thirdly, the subjects of governmentality also participate in its construction, forming individual and collective identities that, in turn, shape governmental activity. Foucault understands governmentality, then, as operating through multiple nodes of transfer, and he stresses that it is not reducible to the state but involves the participation of a heterogeneous range of organisations and individuals such as the church, charitable foundations, philanthropists, doctors and so on. The eugenics movement operated as a supplement to state power in precisely this way, the Eugenics Society fulfilling a particularly important function in this respect. Under the leadership of C. P. Blacker, the Society reconfigured itself after the war, avoiding direct 'propaganda' and seeking instead, in the words of one of its internal reports, to 'exert its influence at the centre by close liaison with cognate organizations' such as the Family Planning Association.[10] While continuing to act as a forum for debate, which attracted influential speakers, the Society was also active in commissioning research and establishing links with academics working on various aspects of the health of the population. It thus generated a pool of specialist knowledge on which politicians could draw as they reflected on the ways in which Foucault's 'multiple body' could be (re)configured as the prime object of governmentality.

The lectures published in *The Birth of Biopolitics*, which focus on the connection between biopolitics and liberal forms of governmentality, are especially relevant to the argument of this book. While analysing the economy of post-war Germany, Foucault also examines the contrast between the nature of post-war reconstruction in Germany (by which he means West Germany) and in Britain. In particular, he discusses the hostile reactions of the 'Ordoliberal' economists Wilhelm Röpke and F. H. Hayek to the Beveridge Report.[11] Röpke argued that the Beveridge plan would lead to 'ever more concentration of power, national income, and responsibility in the hands of the state'; in other words, it was a blueprint for a socialist state.[12] In contrast, Friedrich Hayek, then working at the London School of Economics, warned that 'it is Germany whose fate we are in some danger of repeating', by which he meant that the social structure envisaged by Beveridge was dangerously close to the centrist tendencies of pre-war Germany, which in turn engendered Nazism (189–90). Bizarrely, then, Beveridge is aligned on the one hand with state socialism, on the other with Nazi Germany. As the strength of these reactions suggests, the issue of social security touched on a deep nerve, which is bound up with the question of how far the state should be responsible for and/or have control over the health and vitality of the population. As Foucault argues, it is also intimately connected with the contract between the individual and the state at a time of war, a question he opens up to further analysis by introducing the concept of 'pacts of war' ('les pactes de la guerre').[13] The Second World War, he argues, marks the first occasion when nations waged war not only on the basis of alliances between international partners but also on the basis of social pacts of a rather different kind:

> that promised—to those who were asked to go to war and get themselves killed—a certain type of economic and social organization which assured security (of employment, with regard to illness and other kinds of risk, and at the level of retirement): they were pacts of security at the moment of a demand for war. (216)

Or, as he also puts it, the government was in effect saying to the population: 'Now we are asking you to get yourselves killed, but we promise you that when you have done this, you will keep your jobs until the end of your lives', a formulation that echoes his description of the 'paroxysmal logic' of Nazi eugenics, suggesting that much is at stake in this contract between the individual and the state. It is not exactly a Faustian pact, but it is one in which the risk of the sacrifice of individual life in war is matched by the promise of protection against other risks (e.g. illness) at the level of the population. Individuals are invited to think of their own life as part of a larger whole and to accept that their death (if it happens) will in some sense be compensated for by the lives of future generations. Foucault's use of the term 'pact' here is misleading, suggesting as it does a more-or-less conscious agreement between two groups; the further implication is that this pact was implemented as a coherent biopolitical strategy, whereas as Foucault himself emphasises elsewhere, as a modality of power, biopolitics operates in an uneven, discontinuous and unpredictable fashion, in relation to diverse, often contradictory contexts and pressures.[14] Nonetheless, he here draws sharply into view the eugenic logic that informed much of the planning for post-war reconstruction, and which is made explicit in many of the texts discussed in this book.[15]

NATURE AND NURTURE

The assumption that we can distinguish between nature and nurture as causes of human characteristics is crucial to eugenic thought and structures many of the debates analysed in this study. As Evelyn Fox Keller has argued, in relation to the understanding of heredity the distinction has its origins in Galton's interpretation of Darwin's thinking in the 1860s and 1870s.[16] Darwin speculated that characteristics were handed on through an internal, biological mechanism whereby 'minute granules' (or 'gemmules') were thrown off by parts of the body and collected in the germ cells, from where they were passed on to the next generation. In developing this theory of 'particulate inheritance', he remained open to the idea, first advanced by the zoologist Jean Baptiste de Lamarck, that the environment could influence inheritance and thus the 'gemmules' might be affected by the experience of the organism.[17] However, as Fox Keller emphasises, Galton took a different view, arguing first that the units of inheritance were fixed, and secondly that nature and nurture (or heredity and the environment) must be seen as entirely distinct causal factors in development. In a close reading of a passage from *English Men of Science: Their Nature and Nurture*, Fox Keller highlights the way in which Galton turns the traditional distinction between pre- and post-natal influences into a 'substantive disjunction', a division between two different kinds of 'elements of which personality is composed' (23). The division enabled him to model inheritance as running along two separate tracks with little interaction between them, which in turn made possible his privileging of heredity as the more important factor in human development, as when he claims in *English Men of Science* that 'when nature and nurture compete for supremacy . . . the former proves the stronger', before going on to insist that the most desirable characteristics ('all that has gone into the making of eminent men') are due to heredity.[18] This belief in the determining power of heredity was the main tenet of orthodox or 'mainline' eugenics in the first two decades of the twentieth-century. In the 1930s, however, the movement was taken in a new direction with the emergence of what Daniel Kevles has called 'reform eugenics'.[19] Reform eugenicists reacted against what they saw as the class bias of mainline eugenics and also called for a better integration of eugenic thought with advances in genetic science. As Kevles points out, they were as convinced as the 'mainliners' that human beings were born with inherent qualities, but they were also aware that the mechanisms of genetic inheritance were turning out to be more complicated than had previously been thought. Mendel's work had provided the initial basis for what Julian Huxley called the 'Modern Synthesis' of evolutionary theory and genetics, but it had become clear that inheritance did not proceed through the straightforward transmission of Mendelian 'unit characters' from parent to child. In other words, like did not breed like, as the early eugenicists had claimed. Inheritance was 'polygenic' (multiple genes were involved in the production of characteristics even as apparently simple as eye colour) and extremely complex, so that as Kevles explains, '[i]n human as in virtually all forms of sexual reproduction, genes from one partner were sorted, then combined with those from the other in an infinite variety of unpredictable ways' (146). It was clear that Galton's 'science of breeding' was not going to be implemented in the near future.

Reform eugenicists not only had a more sophisticated understanding of genetics than their predecessors but were also aware that, as yet, little was actually known about the effect of the environment on human development. This became an acute issue during the Depression years, when as Pat Thane has argued, the extent of deprivation among the poor was difficult for the prosperous to ignore, particularly when the unemployed made their distress felt with 'tragic marches of gaunt men through southern towns *en route* to London', providing 'endless grim copy' for radio, newspapers and cinema newsreels.[20] Reform eugenicists like Huxley recognised that poor housing and nutrition, lack of access to medical care and an education that stopped for most children at the age of 14 were likely to be significant factors in the (allegedly) lower levels of 'ability' and 'energy' among the working-classes. He therefore suggested in a lecture to the Eugenics Society in 1936 that little could be done in terms of practical eugenics 'until we have more or less equalized the environmental opportunities of all classes and types—and this must be by levelling *up*'.[21] The concept of 'levelling up' was important for eugenicists who were on the political left, as it enabled them to reconcile the apparently contradictory principles of eugenics and social justice. Richard Titmuss, eugenicist and 'high priest' of the Welfare State, is an exemplary figure in this respect. Titmuss had left school at 14 to work as an insurance clerk and joined the Eugenics Society in 1937 because of his interest in the statistics of fertility. He was active in the Eugenics Society during the war, when he was a member of the society's emergency Committee, and was acting editor of the *Eugenics Review*; he then moved to the London School of Economics to take up a chair in Social Administration and became an influential advisor to the post-war Labour government. Titmuss adroitly reconciled the perspectives of eugenics and social justice in his first book, *Poverty and Population: A Factual Study of Social Waste.* This book opens with the claim that the dwindling numbers of the middle-classes are a threat to democracy, as 'our numbers are being increasingly disproportionately recruited from amongst the unemployed, the unhealthy and mal-nourished, the mentally sub-normal, and from the ranks of those reared in the most unsatisfactory environment and subjected to social and economic hardships'.[22] It thus opens by making a familiar eugenic point, but rather than simply arguing that the answer lies in more middle-class babies, Titmuss makes a powerful plea for saving the lives of 'those children and mothers we already have', those whose poor health is caused by their unsatisfactory environment (301). He argues that socio-medical interventions would save numerous lives, particularly among children living in poverty; they would also improve physical health and mental ability across the board—he points out that the most basic measures, such as providing good breakfasts, have been shown to improve 'mental output and school progress' among disadvantaged children. Titmuss thus makes the case for the alleviation of poverty and sickness on the grounds of social justice *and* eugenic benefit. The equalisation of life chances is a matter of fairness, and at the same time, levelling up will ensure that all those who possess innate 'ability' and 'energy' will be able to fulfil their potential, thus creating a richer pool of 'talent' from which to select 'stock' over the longer term.

When reform eugenicists like Titmuss were discussing the importance of the environment, what they meant was the external environment, that is, the environment outside the body. However, the distinction between heredity and the environment has recently been given a new twist through debates about the relative importance of genetic material and the cellular 'environment' (in other words, the non-genetic components of the individual cell). This follows decades that have been dominated by a narrative of 'biological' or 'genetic' determinism in which genes have been invoked as the determining causes of almost every aspect of human behaviour. The rhetoric deployed by Watson and Crick to describe the significance of DNA and RNA certainly played its part in the construction of this genetic determinism; their initial claim that they had found 'the secret of life' was followed by Crick's articulation of a 'Central Dogma' of genetics, which was that DNA codes for RNA, which codes for protein in a strictly one-way process.[23] This model has now been shown to be too simplistic, but, as Lily E. Kay has remarked, Crick's statement encapsulated 'the imperative logic of the genetic code and the ideology and experimental mandate of the new biology: genetic information, qua DNA, was both the origin and universal agent of all life'.[24] However, as Evelyn Fox Keller and others have argued, recent research in epigenetics has offered a powerful challenge to this model.[25] The term epigenetics was coined by C. H. Waddington as long ago as 1942 to refer to the non-genetic factors in inheritance, but it has taken on an entirely new significance with the discovery that changes in epigenetic factors (in other words, factors over and above the fixed genetic components of the cell) are not only passed on in cell division, but are also transmitted through the generations. Genes and inheritance can thus no longer be considered to be virtually synonymous, and a new model of development is emerging in which, in Fox Keller's words, the interplay among DNA, RNA and the cellular environment is understood to be immensely 'complex and entangled, involving causal influences that extend upward, downward, and sideways' (51). Indeed, the entanglement of factors is so intricate that it no longer makes any sense, she suggests, to attempt to distinguish between heredity and environment, or nature and nurture, as factors in development. Whether we will be able to do without this distinction must remain an open question, but the work of Kevles and Fox Keller is salutary in drawing attention to the contingency and historical specificity of the meanings of 'nature' and 'nurture'.

EUGENIC CULTURES

Drawing on Foucault's analysis of biopolitics and on the work of historians of science and medicine, this book offers a new perspective by examining the broader cultural impact of eugenics in post-war Britain. 'Culture' is as shifting and mutable a term as 'nature', but it is here understood in the sense in which it is widely used in cultural history, to refer to symbolic representations of experience as opposed to empirically verifiable events. It is important to examine eugenics from this point of view precisely because eugenics is not (despite Galton's initial ambitions) a science, but a social and cultural movement that has derived much of its power from its dissemination across a range of discursive fields. Accordingly, this book explores the impact of eugenic ideas on debates in the specialised fields of medicine, sociology, social policy and genetic science. In addition, however, it assesses the impact of eugenic ideas (and ideals) on the general public, examining, among other sources, selected texts from the plethora of paperbacks on such subjects as genetics, evolutionary theory, psychology and social science that were published in this period. Most appeared under the imprint of Pelican Books and were the post-war analogues of the inter-war 'Thinker's Library'; they were written by academics but designed for the general reader and sold in large numbers.[26] The book also focuses on the extensive representations of eugenic thought in literature, not only in 'high' cultural texts such as Doris Lessing's *The Four-Gated City* and *Briefing for a Descent into Hell*, but in popular detective fiction and in science fiction. Eugenic ideas circulate across these discursive fields in complex and sometimes unpredictable ways; hence the book is organised not chronologically but around five overlapping themes.

The first chapter examines the idea of the 'meritocracy', taking as its starting point Michael Young's contention that this should be understood as in part a eugenic measure, designed to 'refresh' the pool of talent of the hereditary elite. His dystopian satire *The Rise of the Meritocracy* is read in concert with a number of other texts in the then-burgeoning field of cultural studies, including Raymond Williams' *Culture and Society*. This chapter also takes in a number of fictions that engage with inheritance and class mobility in varying and conflicted ways, including literary texts such as Williams' own *Border Country* and John Braine's *Room at the Top*. The gender-specific aspects of the meritocracy have to date been almost entirely neglected, but they are situated here in the context of such influential (and eugenically inspired) studies as Myrdal and Klein's *Women's Two Roles: Home and Work*, and they are explored through fictional and autobiographical texts including Margaret Drabble's *Jerusalem the Golden* and Lorna Sage's *Bad Blood*. The second chapter looks at a related issue, that of fears about a fall in 'national intelligence', which was thought to be connected with a higher birth-rate among less intelligent women. The chapter assesses a range of representations of the 'mentally defective' in medical texts and social policy documents and explores the perceived connection between low intelligence and delinquency; it also explores the myth of female promiscuity as both cause and effect of mental deficiency. It analyses representations of the 'defective human' in the context of anxieties about degeneration and post-atomic mutation, focusing on Anthony Burgess's *A Clockwork Orange* and J. G. Ballard's *The Drought*. The close association between social class and mental deficiency is explored with specific reference to C. P. Blacker's concept of 'problem families' and is also tracked through representations of a 'defective' underclass in detective fiction. The third chapter turns to the symbiotic relationship between genetics and eugenics in the post-war era, offering a detailed account of a controversy over Soviet genetics, the 'Lysenko affair', which is analysed in terms of the ideological investments that shaped the Western preoccupation with Mendelian genetics and the Soviet emphasis on the significance of the environment. John Wyndham's critique of Lysenko's ideas in *The Day of the Triffids* is linked with the eugenic themes of his *The Midwich Cuckoos* and *The Chrysalids*, while the chapter also considers ideas about the development of a 'higher' human race that are explored by a number of writers, including Wyndham. The speculative fiction of Doris Lessing and Naomi Mitchison is especially suggestive in this respect, as both writers engage with the notion that atomic radiation might prompt the emergence of more 'gifted' and 'advanced' human beings.

Chapter 4 analyses the heated debates in this period over the genetic basis of race, focusing on C. D. Darlington's contentious argument that mixed race individuals ('hybrids') were biologically inferior and on the Eugenics Society's 1958 attack on West Indian immigration, which similarly expressed concerns about the dangers of miscegenation. The chapter also assesses attempts by social anthropologists such as Sheila Patterson to present race as a social construction rather than a biological category: these were only partially successful, and the eugenic undertow in texts such as Patterson's *Dark Strangers: A Study of West Indians in London* is discussed. Representations of immigrants in novels such as George Lamming's *The Emigrants* and Sam Selvon's *The Lonely Londoners* are considered alongside those in the fiction of Muriel Spark and Colin MacInnes. The fifth chapter considers the ways in which the concerns of British eugenicists (which were intimately bound up with a privileging of 'Anglo-Saxon stock') were projected onto the world stage. It assesses the extent to which ideas about 'inferior' races inflected debates about over-population, and it focuses in particular on C. P. Blacker's role as chair of the International Planned Parenthood Federation (IPPF), which had its headquarters in London. The IPPF played a prominent part in one of the worst excesses of the 'population movement', the sterilisation campaign in India in the 1970s, which is here related to its geo-political contexts and 'read back' through Salman Rushdie's critique of Indira Ghandi's policies in *Midnight's Children*. Finally, this chapter explores the convergence of British anxieties about populating the empire and the Australian government's 'White Australia' policy. This led to the assisted passage scheme under which more than a million 'ten pound poms' emigrated to Australia between 1947 and 1972. This chapter explores the complex intentions and effects of this scheme, reading government policy documents in conjunction with the 'migrant' writer Elizabeth Jolley's fictional autobiography *The Vera Wright Trilogy*. The afterword explores the connections between post-war eugenics and contemporary 'liberal eugenics', tracing the continuities between current developments and the eugenic assumptions that prevailed in post-war Britain.

1 Eugenics and the Meritocracy

> The highest fulfilment lies in submission. Of nothing is this more true than of society; and here no lesson has been more simple, and yet more painful than the fact of genetic inequality.[1]

> It was a considerable achievement for a society to pour so much milk and so much orange juice, so many vitamins, down the throats of its children . . . in spite of the fact that the statistics of healthy and intelligent childhood were stretched along the curve of achievement, and only a few were allowed to travel through the narrow gate at the age of eleven, towards the golden city.[2]

Meritocracies are generally thought to be self-evidently good things. In Britain, a society in which merit is rewarded regardless of background or social class has long been promoted as an ideal by politicians on both the left and right, together with the closely linked concepts of equality of opportunity and social mobility. Yet when Michael Young first coined the word in *The Rise of the Meritocracy*, it was meant as a pejorative term, as shorthand for the technocratic, deeply divisive society that he thought would result from post-war changes in education.[3] Young saw the selective system enshrined in the 1944 Education Act as a biopolitical strategy designed to maximise national 'talent' at a time of increased international competition; he also linked it with the focus on intelligence (as opposed to physical strength or mental health), which was an especially marked feature of British eugenic thought. This chapter traces the connections between eugenic thought and educational policy in the years leading up to the war, before assessing the long term implications of the 1944 Act, which many perceived to be eugenic both in intention and effect; certainly, both supporters and critics of selective education saw it as marking a significant shift in the way that the relationship between the individual and the state was configured. Political and social debates about the issue are here considered alongside a number of literary texts that shed light on the subjective and affective dimensions of upward social mobility in this period.

The 1944 Act was steered through Parliament by the progressive Conservative Education Secretary R. A. Butler but was the culmination of long-drawn-out debates over the expansion of secondary education. The process started with the Education Act of 1918, which introduced a school leaving age of 14 and required all elementary schools (the state schools attended by the overwhelming majority) to provide more advanced courses for their older pupils. The Hadow Report of 1926 recommended a break between primary and secondary education at the age of 11 and, for the first time, the division of secondary schools into three types—grammar, technical and modern. These recommendations were carried out to a limited extent, if there were sufficient local authority funds. The Spens Committee, reporting in 1938, endorsed the Hadow Report and strongly supported IQ tests as the best means of selecting children for ability at age 11. The case for selective education was further strengthened by the 1943 Norwood Report, chaired by a former headmaster of the Harrow School. This claimed, though without offering any evidence to support the claim, that there existed three 'rough groupings' of children who had 'different types of mind'—those interested in learning for its own sake, those whose abilities lay in the field of applied science and those with more 'concrete' skills.[4] Evidence was submitted to this committee by a wide range of employers and teachers' associations, almost all of whom accepted the validity of this crude system of classification. Only the Trades Union Congress (TUC) (which represented rank and file workers), in a response to the Spens Report included in the evidence to the Norwood Committee, protested over the social implications of such a hierarchical system. Their comments are worth quoting, as they emphasise the life-long effects of educational segregation:

> The separation of the three types of school is . . . also bound to perpetuate the classification of children into industrial as well as social strata. So long as the three types of school remain separate, it is inevitable that the Grammar School pupil will continue to look upon the black-coated job as his natural right; that the Technical High School will tend to be regarded as the training ground of foremen and of highly skilled workers in the engineering and building trades; while pupils leaving Modern Schools will have to be content with what jobs are left.[5]

The TUC was also virtually alone at this point in arguing for 'multilateral' (i.e. comprehensive) schools.

The expansion of secondary education was supported by the Labour Party on the grounds of social justice. It was widely recognised that the education of most children beyond the age of 11 was inadequate, and it was also known that more than half the working-class children selected for free places in grammar schools turned them down, as parents could not afford the loss of wages for an extra two years. Labour politicians were anxious about the generally low level of education but also felt that it was particularly unfair to condemn talented children to elementary or 'modern' schools. It was on this issue that their interests converged with those of eugenicists who had become concerned about the prospect of a fall in 'national intelligence'. This question was drawn to the attention of the wider public with the publication of Raymond Cattell's polemic *The Fight for Our National Intelligence* in 1937. Cattell was a psychologist whose research had been funded by the Eugenics Society, and in the first chapter of his book, he made the sensational claim that:

> every clinical psychologist knows, with the conviction of 100 to 1 probability, that national intelligence is falling disastrously at the present moment through the continual rapid replacement of the constitutionally bright by innately dull and limited types.[6]

Cattell had worked with the pioneering educational psychologist Cyril Burt, who not only endorsed Cattell's views about the decline in intelligence but later gave evidence on this topic to the 1949 Royal Commission on Population, estimating a future decline of about 1.5 points of IQ over each generation. Burt's estimates had a tangible effect: the Royal Commission accepted his view and accordingly recommended the introduction of adjustments to the tax system in order to encourage the 'educated classes' to have more children.[7]

Burt was closely involved with the Eugenics Society, and his work was a central point of reference for eugenicists for more than four decades. His interest in the subject stemmed from an early connection with Francis Galton, whom he met through his country doctor father. After training in psychology and physiology, Burt began a study comparing the intelligence of boys from a preparatory school and an elementary school in Oxford and a school in the Liverpool slums. Most of the children in the first group were the sons of 'academics, Fellows of the Royal Society, or bishops', in other words, they came from just the kind of family on which Galton had based his arguments for 'hereditary genius'.[8] Burt, accordingly, came to the same conclusion as Galton—that intelligence was strongly inherited. He maintained this position throughout his career, to the extent of falsifying the data in papers he published in the 1950s and 1960s on intelligence in identical twins.[9] His biographer L. S. Hearnshaw has suggested that these papers (which appeared after his retirement) were written at a time when Burt was feeling particularly embattled, as the tide was turning against an emphasis on heredity, and indeed against the very conception of IQ as a single, quantifiable entity.[10] However, the deception was not revealed until 1974, and from the 1920s to the 1960s, Burt was a powerful figure, one of the experts on whom government could call in order to shape what Foucault would term governmental reason. He became an advisor to the Home Office, the Ministry of Health and the Board of Education, and he is cited by Greta Jones as a pre-eminent example of the 'technicians of social engineering' who came into their own with the expansion of social welfare during and after the Second World War.[11]

Through his work on IQ tests, Burt also provided an indispensable instrument of governmentality. IQ tests had first been developed by Alfred Binet at the request of the French government, in order to detect 'mentally deficient' children. Binet drew up a series of tests designed to measure the qualities of memory, reasoning and verbal ability, and with his colleague Théodore Simon, he developed the concept of a quantifiable mental age. In 1913, Burt was appointed to the London County Council to test for 'defective' children so that they could be transferred into special schools. As Muzumdar points out, although he was supposed to be testing only subnormal children, by 1915 he had set up general research into the distribution of intelligence and was refining the design of standard IQ tests.[12] He was also inspired by the use of IQ tests by the US Army in 1917, in which military recruits were tested for intelligence and aptitude before being assigned to particular tasks. Seeing this as a model that could be applied to the civilian population, he began to campaign for the use of IQ tests to identify bright children from all social classes, whose intelligence, he argued, should be viewed as a national resource. In 1924, he assured the Hadow Committee that it was possible to make an accurate assessment of a child's mental ability by the age of 12, and his support for IQ tests carried great weight with successive governments.

In the inter-war years debates about education were framed in terms of the competing discourses of social efficiency and social justice. The perspective of social efficiency was invoked both by those who argued that the low level of education for the poor was leading to a serious waste of national talent, and by those who took the view that as intelligence was genetically determined and already concentrated in the higher social classes, there was little point in wasting time and money seeking it out amongst the poor. The case for moving towards greater social justice through education was urged by socialists such as the historian R. H. Tawney, but only up to a point. In his Labour Party pamphlet 'Secondary Education for All', for example, he argued that all children should be educated to the highest possible standard, but that as individual needs and abilities varied, secondary schools should be of different types.[13] The 1944 Act, which like the Beveridge Report is often seen as a touchstone of post-war optimism and egalitarianism, was inclined more towards the principle of social efficiency than social justice. It was progressive in making state education free to all and in raising the school leaving age to 15, but it was reactionary in its endorsement of the tripartite system of secondary education, through grammar, technical and modern schools. Moreover, a lack of investment in technical schools meant that the system soon became, in effect, a binary one, in which 20–25% of the school population went to grammar and the rest to secondary modern schools (Figures 1.1 and 1.2).

Figure 1.1 Manchester Grammar School, 1950, morning prayer.

Figure 1.2 Bourne Secondary Modern School, Ruislip 1950, handicraft class.

LADDERS OF OPPORTUNITY

The Rise of the Meritocracy offers a scathing critique of this educational system and of the social forces that produced it. The argument is conducted through a fictional essay, supposedly written in 2033, on the May risings associated with the 'Populist Movement', an alliance of women and working-class men who have joined together to protest against the rigid social order of the twenty-first century. Young extrapolates from mid-twentieth-century trends to imagine a society geared towards social efficiency, which has massively extended the selection processes developed during the two world wars. The Second World War, Young's essayist writes, 'woke people up to the fact that the nation possessed a supply of ability never ordinarily used to the full'; he also notes that international competition was a further spur for the introduction of competitive selection across all aspects of social and economic life.[14] In this society an aristocracy of talent has replaced an aristocracy of birth, with rule 'not so much by the people as by the cleverest people'. And as its populist critics argue, 'intelligence' is in this context defined solely in relation to the instrumental criteria of productivity and economic expansion. Thus:

> People are judged according to the single test of how much they increase production, or the knowledge that will, directly or indirectly, lead to that consummation. . . . The ability to raise production, directly or indirectly, is known as 'intelligence': this iron measure is the judgement of society upon its members. (167–8)

In addition, Young highlights the biopolitical and eugenic effects of a move to a society in which intelligence and social status have become inextricably linked. His essayist cites Galton as the main inspiration for the new social order—his 'Oh God, oh Galton!' being a clear echo of Huxley's 'Our Ford' in *Brave New World*—and frequently refers to the Fabian eugenicists Shaw and Beatrice Webb as guiding spirits. For in this dystopia the longstanding eugenicist concern with IQ has been taken to its logical conclusion. This is a society in which classes are 'biologically' differentiated according to merit: each of the components of merit—IQ and a propensity for effort—is thought to be quantifiable and identifiable by psychometric testing. Accordingly, the 'talented' are assigned to the 'level which accords with their capacities', so that over time the ranks of the gifted 'have been swelled, their education shaped to their high genetic destiny' (15). Moreover, the elite is more or less unassailable because it is legitimated by the biological 'truth' of genetic endowment:

> Today all persons, however humble, know they have had every chance. They are tested again and again. . . . But if they have been labelled 'dunce' repeatedly they cannot any longer pretend; their

> image of themselves is more nearly a true, unflattering, reflection. Are they not bound to recognise that they have an inferior status—not as in the past because they were denied opportunity; but because they *are* inferior? (108)

Social stratification is enforced through tight regulation: IQ registers are kept at Eugenics House, and National Intelligence Cards are regularly updated. The drive to maximise 'talent' is supported by the campaign for 'intelligenic marriage' in which people are encouraged to consult the intelligence register before marrying, as meritocrats must not waste their genes. The importance of this issue is signalled through Young's deft re-working of a familiar twentieth-century romance fiction plot, in which an aristocrat seduces a pretty working-class girl but their marriage is blocked, until it turns out—in the nick of time—that she herself is the long-lost daughter of an aristocrat. Young introduces a twenty-first-century version of this theme, in which the 'pretty young mother', who has been rejected with her child not because of her social status but because of her poor intelligence record, finds that 'she is going to be all right after all, the Registry has wrongly docketed her grandfather'—he turns out to have had the high IQ rating, which is the new equivalent of noble blood, transforming her at a stroke into an acceptable marriage partner (174).

As Young makes clear, the aim of the meritocracy is not equality but 'equality of opportunity', or, to put it more accurately, this is the myth on which his meritocracy is founded. Like all such myths it involves paradox: the meritocracy is a society in which it is alleged (by the elite) that everyone is given a fair opportunity to find the role for which he or she is suited. Such an argument conveniently overlooks the fact that in a highly stratified society, in which 'talent' is identified before birth and each child assigned to an appropriate educational stream from the nursery onwards, there is no genuine equality of access to resources and support.

Young's critique exposes a further confusion or blind spot that continued to muddy eugenic arguments. After the war even reform-minded eugenicists such as Blacker continued to invoke the principle of genetic determinism in support of their arguments for eugenic policies. For example, in his 1952 book *Eugenics: Galton and After*, which Blacker saw as offering a 're-orientation' of eugenics for a post-war society, he argues, disingenuously, that because of the complexity of the mechanisms of inheritance in man, we are justified in continuing to act on 'the assumption whereon our ancestors who domesticated animals and plants have always acted, that like produces like'.[15] Young challenged this view, his essayist drawing on the work of Hans Eysenck to argue that there can never be any certainty of the direct transmission of characteristics from parent to child: a child's inheritance is best understood as being transmitted *through* rather than *from* its parents.[16] Thus, clever parents will sometimes have less intelligent children, whom they will want, nonetheless, to have all the advantages they have had. This is one of the factors that leads the 'new Conservatives' in the meritocracy to make their proposal that a privileged education should be *guaranteed* for the children of the elite, without IQ or other testing. The convergence of this development and plans to close the adult education centres that offer 're-testing'—and thus the hope of social mobility—throughout life are the trigger for the successful populist uprising and a twenty-first-century Battle of Peterloo in which the essayist is himself killed. The critical factor in the destruction of the fictional meritocracy thus turns out to be the hopes and ambitions parents have for their children: as the narrator has already pointed out, 'the home is still the most fertile seed-bed of reaction' (31). The pertinence of all this to parents' struggles to get their children into grammar schools—'through the narrow gate to the golden city', as Steedman puts it—is clear.

The Rise of the Meritocracy targeted what it saw as a eugenic re-ordering of society involving the skimming off of working-class talent in order to refresh the elite. Young references *Brave New World* and is clearly drawing on Huxley's vision of a genetically engineered society and suggesting that, even as he wrote, an 'alpha' class was being created in the grammar schools of Britain. He was not alone in reading this aspect of post-war reconstruction as a form of social, or socio-biological, engineering. It is instructive in this respect to read his critique alongside that of the critic and novelist Raymond Williams, who approaches the subject from a different perspective but reaches similar conclusions. Young was a sociologist and policy maker, who drafted the Labour Party manifesto for the 1945 general election and was linked to a number of initiatives of the post-war Labour government.[17] Williams, by contrast, was a literary critic and, unlike the privately educated Young, was himself a meritocrat, the son of a railway worker and beneficiary of the more limited free educational opportunities that had existed between the wars. Central to Williams' critique of the meritocracy is a very specific understanding of the working-class culture in which he grew up, one alert to its 'structures of feeling', the celebrated phrase he used in *The Long Revolution* to denote the 'felt sense of the quality of life at a particular place and time'.[18] Williams associates this culture with communism, socialism and co-operation; in a word, with 'solidarity'. In *Culture and Society*, he argues that in working-class culture, 'improvement is sought, not in the opportunity to escape from one's class, or to make a career, but in the general and controlled advance of all'.[19] It is for this reason that Williams opposes the principle of the meritocracy: it undermines solidarity. For Williams, the 'ladders of opportunity' that it offers are 'a symbol of the bourgeois idea of society' (i.e. a belief in individual self-advancement). The crucial point about a ladder, he points out, is that you can only climb it alone. So, he argues:

> the ladder version of society is objectionable in two related respects: first, that it weakens the principle of common betterment, which ought

> to be an absolute value; second, that it sweetens the poison of hierarchy, in particular by offering the hierarchy of merit as a thing different in kind from the hierarchy of money or of birth. (317–18)

This passage occurs towards the end of Williams' wide-ranging analysis in *Culture and Society* of the interactions among industrial development, class, democracy, art and culture over the period since the Industrial Revolution. What distinguishes Williams' argument and approach throughout the book is his focus on individual perception and understanding as active agents of social transformation or, to put it another way, his emphasis on the ways in which mentalities, assumptions and feelings, as well as ideas and/or political actions, are deeply implicated in the processes of social change. This is, perhaps, Williams' most significant contribution to social and cultural thought.

In *Border Country* (1960), Williams puts this theory into practice, using the novel form to explore the subjective experience of those who are caught up in social change. The novel's project and its rationale is suggested at the outset as Matthew Price, a lecturer in history at London University, ponders the difficulties of his research on nineteenth-century population movements in the Welsh mining valleys. He has completed the empirical part of his work but is struggling to capture the subjective experiences of the people who moved to the valleys, which can only be accessed through letters and diaries. He comments that in order to capture the 'change of substance' involved in social mobility, he needs a more intuitive approach to an experience that 'I can feel but not handle, touch but not grasp'.[20] Williams' aim in this novel is to provide the kind of record that is lacking in relation to the population movement Matthew is investigating and to capture the complex movements of thought and feeling associated with the experience of post-war social mobility. Matthew has himself climbed this 'ladder of opportunity', moving from a childhood in the Welsh borders to an education at Cambridge and his current academic position, but he has many unresolved feelings about the cross-class, inter-generational transition he has been through. He is called back to his home when his father has a serious heart attack, and through a series of conversations with his father and reflection on key moments of his past, he begins to come to terms with the traumatic aspects of the dislocation he has experienced.

Matthew's father Harry Price stands as an exemplary figure, the embodiment of the 'solidarity' and commitment to the common good that Williams associates with traditional working-class culture. He has fought in the First World War and during the General Strike of 1926 comes out with his colleagues in support of the miners; moreover, when his closest friend Morgan starts up a business, he refuses an invitation to join him and steadfastly rejects subsequent requests because he sees business (or 'dealing') as a form of exploitation. Morgan feels this refusal keenly because of his admiration for Harry, which comes out in a conversation with Matthew towards the end of the novel. For Morgan, Harry is an exceptional man because he has shaped his values in relation to his own culture. As Morgan puts it, '[t]he mind he's got is to the things we say really matter. We say it, and run off in the opposite direction' (358). Both Morgan and Matthew have been driven by the desire to get ahead, and this has created a distance between Matthew and his father that comes out in the last conversation between them. Harry has wanted Matthew's success but is wary of its exclusionary aspects, and he warns his son that a life's work should involve more than individual achievement:

> 'There's no virtue in the work, but that men should stand as they are.'
> 'Stand equal?'
> 'Stand as they are, with nothing bearing them down. For you that was made quick.' [i.e., quickly]
> 'Part of it was made quick.'
> 'Only it isn't solved, when it's made quick for you. The rest of us need it, remember.' (388)

The relationship between the two is one of more than usual ambivalence, and some of the causes of this are suggested through the undercurrent of violence that runs through the text. The narrative is structured around the stages of Harry's illness, represented as a series of assaults on his bodily integrity, and his somatic decline is vividly and insistently rendered. Paradoxically, at the moment when Harry has his final heart attack, Matthew (then on his way back to London) goes through a telepathic experience in which he identifies with and feels almost overcome by the bodily strength of his father, who has also become one with the body of the people:

> Closing his eyes, he saw Harry's heavy body, and the crowd moved in it, the crowd in its constant pressures. Through his whole body he could hear the deep, strong voice, and the rhythms went out into all the voices around him, until he heard his own voice, differently pitched. For some moments this lasted, the voices rising, until it seemed they would break him. (393)

In this moment, Matthew metaphorically struggles with, and eventually overcomes, his father and the 'ordinary' people around him. It is as though they must be abjected in order that he—and others climbing the 'ladder of opportunity'—can go forwards with increased vigour. And Matthew is revivified by his father's death; his wife notices 'that he was evidently stronger, that certain energies were released. There was a difference of physical presence' (435). This outcome resonates strongly with Foucault's account of the internal logic of biological racism, experienced as the sense that 'the more inferior species die out . . . the stronger I will be, the more vigorous I will be'.[21]

Probably the best-known literary exploration of the theme of social mobility is John Braine's *Room at the Top*, which is set in different geographical territory in the industrial north of England. This novel's title rapidly passed into the language, so that, for example, in a 1960 essay on the new managerial elite, the Labour politician Peter Shore identified the idea of the 'room at the top' as 'the dominant obsession and the compelling myth of our time'.[22] Braine's novel is not an endorsement but a critique of this obsession and its potentially damaging consequences. *Room at the Top* tracks the social trajectory of Joe Lampton, a working-class grammar school boy who (like Williams and his protagonist Matthew Price and Braine himself) has profited from the inter-war 'ladder of opportunity'. The society Braine evokes is one of competitive individualism, endorsed by all (except the retrospectively critical first-person narrator) because it fits the needs of the post-war economy. The text is suffused by the language of economics; Joe is a local government accountant who spends his evenings studying the theory of surplus value—though whether with a view to Marxist critique or to more efficient exploitation of capitalist principles remains unclear. When he starts a relationship with the young and innocent Susan, who is also a wealthy heiress, his inability to reciprocate her love is presented in terms of a failure 'to accumulate enough emotional capital' to match 'her gift'.[23] As the term 'emotional capital' suggests, for Joe almost every aspect of life depends on economic capital and everything is reducible to the principles of economic exchange. So, for example, in a sly parody of local authority bureaucracy, he and his friend Charles have written up the strong correlations they have observed between a husband's income and the attractiveness of his wife:

> We even typed out a schedule, the Lampton-Lufford Report on Love. There was an appendix with Sex Summaries. . . .
>
> The grades corresponded, naturally, with the income of husband or fiancé, running from One, for millionaires and film stars and dictators—anyone with an income over £20,000 in fact—to Twelve for those under £350 and not likely to get any more. Charles and I belonged to Grade Seven. (37)

In these terms, Joe's failure to develop the requisite 'emotional capital' for Susan can be causally related to his lack of economic capital: as a 'Grade Seven', he seems to have no hope of acquiring a 'Grade Two' wife. However, in another, gendered twist on the stereotypical romance plot, whereby a penniless young woman marries an aristocrat, Joe does succeed in rising to 'the top', and he does this very much on the basis of 'merit'. He is described as a bright lad who has used the time when he was held in Stalag 1000 as a prisoner of war to study for his accountancy exams. Possessing 'IQ + effort', he has thus rapidly moved from the working-class to the white-collar class. He also quickly realises that he must have what

Bourdieu would call 'cultural capital' as well as intelligence if he is to succeed further, and as soon as he moves to a new job in Warley, he joins the local theatre group and attempts to develop what he thinks of as the more general 'polish of manner' necessary for social success.[24] However, Joe's rise to 'the top' owes even more to his biological capital. There are numerous references in the novel to his physical strength and his attractiveness to women. Ruthlessly Joe exploits these attributes in his relationship with Susan, alternating between tender words and 'rough' physical treatment of her so that she is caught up in a pattern of longing for him—very much as the pretty young girl might exploit her charms in a more conventionally gendered romance. Most significant, however, is the fact that Joe finally achieves his goal through making Susan pregnant, thus forcing their marriage and his assimilation into her father's business empire.

Joe's cross-class marriage is of a type represented in many other literary texts of this period: for example, *That Uncertain Feeling*, *Look Back in Anger* and *A Kind of Loving*. In each of these texts, a young working-class man marries into a higher social class, in a social pattern dubbed 'male hypergamy' in an article by the influential social anthropologist Geoffrey Gorer. He made the point that in Britain, female hypergamy (marrying upwards) was a well-established and relatively common occurrence; alluding like Michael Young to the structures of romance fiction, he notes that it has 'informed the dreams of any number of unprivileged girls, from *Jane Eyre* to most of the serials in most of the current women's magazines'.[25] However, male hypergamy was a new phenomenon, brought about because the 1944 Education Act had made it possible for 'bright young working-class lads' to jump three or four social classes in the space of a decade (246). According to Gorer, this inevitably produced conflict, because although these young men had a changed intellectual outlook, they remained trapped in the gender roles of their class of origin. Working-class men, he claimed, identified with a role that involved the significant display of physical force and the expectation of subservience from a wife, whereas upper-middle-class women expected to be treated as equals and to be shown consideration and respect. Braine's novel suggests, however, that the reality was more complex. In both *Room at the Top* and its sequel *Life at the Top*, Joe exploits his physical attractiveness but feels that he, in turn, is being exploited because of it. His identity has become primarily a bodily identity: he is described in terms of meat on more than one occasion, the friend of his lover Alice, for example, commenting, 'Ah, I don't blame Alice. You're the sort of man I like, big and beefy'. The identification of Joe with the somatic has a doubly objectifying effect. It reverses gender roles, feminizing him by rendering him the object rather than the subject of desire; it also de-individualises him and underscores the extent to which his social rise is part of a wider movement within the population.

What Braine also highlights, however, is the significance of reproduction as a flashpoint for anxieties surrounding cross-class mobility. Reproduction is always a public as much as a private matter, laden with social and economic meanings and closely bound up with hopes and expectations for the future. In particular, as the anthropologist Marilyn Strathern has put it, in most cultures of the world when a child is born, 'what is reproduced is a set of social relations'.[26] Reproduction across class boundaries is often perceived as a significant threat to existing social relations, as it manifestly dissolves the apparently fixed boundaries between classes. In mid-twentieth-century Britain, these boundaries were frequently shored up by an appeal to 'biological' or 'natural' differences between the classes. So in *Room at the Top*, the ambiguous trope of 'breeding' is used to denote the quality possessed by Joe's rival for Susan's affections, the mustachioed ex-RAF officer Jack Wales. The ambiguity lies in the fact that breeding can refer either to biological or cultural inheritance, birth or training, but in this context, it is clearly meant to imply belonging to a biologically distinct class. The even more specific term 'caste', with its quasi-racial implications, is applied to Susan's mother, the last of the aristocratic St Clair family. A further key trope in this respect is that of blood, good as opposed to bad, the blue-bloodedness of the aristocracy as opposed to the raw red-bloodedness of the proletariat. This powerful nexus of metaphors suggests the anxieties surrounding cross-class reproduction, which is represented as being almost as dangerous as miscegenation—indeed it is explicitly presented in these terms in Marghanita Laski's 1952 novel *The Village*. The relative reproductive failure of cross-class marriage as represented in a range of literary texts is a further index of this fear. So in *Life at the Top*, for example, we learn that Joe's second child, Barbara, is not his: Susan has cuckolded him with an old family friend, someone who is 'whatever happens . . . a gentleman'—unlike Joe (230).

These themes are even more prominent in John Osborne's play *Look Back in Anger*, the *locus classicus* of the representation of both the 'Angry Young Man' and male hypergamy. Jimmy Porter, working-class but educated in one of the new 'white tile' universities, has married the 'well-bred' daughter of an ex-Indian Army colonel. While Alison's pregnancy and miscarriage are not actually seen in the play, they are at its heart and constitute the most significant 'action' within it. Long before Jimmy knows of Alison's pregnancy, he wishes a miscarriage on her while at the same time identifying himself as her foetus and Alison as a figure of the monstrous-feminine: 'That bulge around her navel—if you're wondering what it is—it's me. Me, buried alive down there, and going mad, smothered in that peaceful looking coil'.[27] Alison is a 'devouring python' who has ingested and destroyed him, the metaphor suggesting not only her malign power but that sex between the two of them may create monsters. Cross-class reproduction is thus refigured as cross-species reproduction, with Jimmy explicitly invoking the trope of monstrosity when he finally learns of Alison's pregnancy: 'I don't care if she's going to have a baby. I don't care if it has two heads!' (73). The metaphor of blood is also deployed, as Jimmy contrasts upper-class blood ('pale Cambridge blue') with working-class blood ('it looks like cochineal, ever so common'). More tellingly still, at a crucial moment in the play, Jimmy represents Alison as battening on his blood:

> Why, why, why, why do we let these women bleed us to death? Have you ever had a letter, and on it is franked 'Please Give Your Blood Generously'? Well, the Postmaster-General does that, on behalf of all the women of the world. I suppose people of our generation aren't able to die for good causes any longer. (84)

Osborne condenses a number of themes here. On one level, Jimmy's comments re-invoke the theme of the re-invigoration of 'thin' aristocratic blood through marriage with someone of the lower classes. What distinguishes the marriage of Jimmy and Alison, however—as with Joe and Susan in *Room at the Top*—is the fact that this is *male* hypergamy, and it is the component of gender reversal that causes Jimmy to feel that he has been un-manned in his relationship with Alison, objectified and inserted into a social and economic order over which he has little or no control. Yet Jimmy's complaint that he is being 'bled to death' by women has a further implication. It is linked with the idea that people of his generation have no good causes to die for anymore, a claim followed almost immediately by these celebrated lines:

> There aren't any good, brave causes left. If the big bang does come, and we all get killed off, it won't be in aid of the old-fashioned, grand design. It'll just be for the Brave New-nothing-very-much-thank-you. (94–95)

The juxtaposition suggests that Jimmy and others like him are indeed 'dying' for a cause, and for a new kind of 'grand design', which is the manipulation of the population in the name of the greater social good. While young men of his generation are not risking death in a struggle against fascism, their lives are nonetheless being 'taken' as the raw material for social planning in the name of national efficiency. The assumption is that once they have been educated and trained, their 'talent' will be assimilated to (or ingested by) the elite. However, Osborne's text suggests that this assimilation may not be as straightforward or painless as might have been thought—particularly as we learn at the end of the play that Alison has not only miscarried but has become infertile. This marriage, then, will not be (re)productive.

GENDER AND THE MERITOCRACY: WOMEN'S TWO ROLES

Almost all discussion of meritocracy and post-war educational change focused then (as, to a large extent, now) on the opportunities being offered to working-class boys.[28] The emphasis on the 'Angry Young Men' is exemplary in this respect. Like others who were around at the time, Doris Lessing has stated firmly that:

> The Angry Young Men was a phenomenon entirely invented by the newspapers, the media. It went rolling on, year after year, gathering momentum, and all the time I was amazed no one seemed to notice that in fact they had very little in common.[29]

Yet despite the protestations of almost all involved, the figure of the angry young man acted as a valuable cultural shorthand, a means of articulating real changes in the perceptions and expectations of young working-class men in the 1950s. Similarly, the focus by social commentators on the dangers of hypergamy, and especially on the conflicts experienced by the 'scholarship boy', provided a form of external legitimation for what would otherwise have remained private anxieties.[30] By contrast, there was very little public discussion of the difficulties faced by the scholarship girl or 'female meritocrat'. The work of Michael Young is an exception, with its anticipation of some of the problems that would confront women in a meritocracy. Young recognised that gifted women would inevitably struggle with what was known at the time as women's 'dual role'. In his fictional meritocracy, women are not expected (or allowed) to make their contribution to national production indefinitely, as they are also expected to make a different kind of contribution, by fulfilling their reproductive role. So, as he writes:

> [T]hey take the post for which they have been trained only until they marry. From that moment they are expected, for a few years at any rate, to devote themselves to their children . . . they cannot, if they take any notice of the teaching of psychology, entrust the entire care of their offspring to a person of low intelligence. Infants need the love of a mother; they also need her intellectual stimulation, her tender introduction to a high culture, her diligent preparation for a dedicated life. (171–72)

Young here captures exactly the contradictory rhetoric that surrounded women's position, expected as they were to exhibit a 'devotion' to their children that would preclude any sustained professional 'dedication' on their own part.

This analysis must be placed in the context of an understanding of 'women's two roles', which was very different from that which prevails today. While it is now assumed that women's dual role involves juggling a career and motherhood simultaneously, in the post-war era, women's dual role was always conceived sequentially. This was the assumption made by Richard Titmuss, for example, when, in a lecture given in 1952, he highlighted women's increased life-expectancy and the fact that they were having smaller families; for this reason, he argued, they had many more 'unencumbered, active years' in which they would be available for work having brought up their children.[31] Alva Myrdal and Viola Klein, in their hugely influential study *Women's Two Roles: Home and Work*, also take these demographic changes as their starting point, but they go further than Titmuss in constructing women's 'two roles' in terms of their duty to the state. This is hardly surprising given Alva Myrdal's connections with the eugenics movement. With her husband Gunner, she was a key figure in the 'scientific' planning of the post-war Welfare State in Sweden, which espoused both positive eugenics (in the form of allowances for mothers) and negative eugenics, including compulsory sterilisation.[32] *Women's Two Roles*, together with texts such as Judith Hubback's *Wives Who Went to College* (1957), can thus be seen as a post-war version of the eugenic feminism first advocated in the 1880s and 1890s.[33] Myrdal and Klein emphasise the absolute priority of women's reproductive role and, in particular, stress that it is women's duty to improve the *quality* of the next generation:

> 'Children First' is the motto writ large over all discussions of the merits and demerits of married women's employment. Our children are our stake in the future; hence their well-being is of vital concern to society as well as to us personally. And as we have brought them into this world, it is our responsibility to make them, as best we can, fit to live in it happily and successfully.
>
> The quality of the next generation has become of even greater moment since the size of families has shrunk. With a birth-rate at, or only just above, reproduction level we can less than ever afford the risk of failures. To-day we have learned enough about the formation of personality to understand the great extent to which its development depends on maternal care and affection, particularly during the first years of life.[34]

In relation to the eugenic feminism of an earlier period, it is worth noting, however, that the terms of the argument have shifted significantly in response to social and cultural changes. Unlike earlier middle-class mothers, the post-war generation could no longer expect to have nannies to care for their children; moreover, largely as a result of the psychologist John Bowlby's influential work on the mother–child bond, it was widely thought that children needed the undivided attention of their mothers in the early years of life.[35] So, as Myrdal and Klein make clear, it is now middle-class women's eugenic duty not only to have children but to dedicate themselves to nurturing them in the pre-school years.

On the other hand, Mrydal and Klein take a bracing approach to women's duty to work under all other circumstances. Although the needs of children are seen as paramount, they brush other 'family obligations' aside, construing women's 'readiness not only to change their jobs but to abandon their careers' as 'a problem of social and economic waste' (109). Offering something of a hostage to fortune, Myrdal and Klein go on to raise the issue of the cost of women's education, arguing that:

> If this [the question of social and economic waste] is true of all women, it particularly applies to women in jobs which require a long or expensive training. It is, in fact, quite surprising that parents, employers and education authorities have not been more chary of investing time and money in the higher education or specialized training of girls, in view of the short time most of them are likely to make practical use of the qualifications they have acquired. (109)

Throughout, their discussion of the dual role is couched in terms of the need to deploy women's reproductive and productive power efficiently. While it is possible that their arguments may have been phrased in this way in order to appeal to the economic interests of government, the use of military metaphors in relation to motherhood (with un-ironic references to 'mothers on active service' and 'ex-service mothers') betrays a more deeply entrenched eugenic perspective.

Myrdal and Klein were particularly concerned with the 'educated, intelligent women' who were the target of what might be termed a crypto-eugenic campaign in the post-war years. As noted in the Introduction, the 1949 Royal Commission on Population Report devoted considerable attention to the question of 'differential fertility', that is, the idea that 'educated persons' were having relatively smaller families. This trend, they suggested, was linked to the fact that existing financial arrangements disadvantaged those who had children, a fact of which the more educated (and more prudent) were likely to be well aware. To counter this, the commission recommended increases in family allowances and tax relief across the board, but they also put forward a proposal for special 'professional schemes' of family allowances. These were already in existence in universities, as Hilary Rose has recalled:

> Positive eugenic policies were like the £50 per year added to my salary for each of my children when I was a lecturer at the London School of Economics in the 1960s. The explanation for this largesse was that William Beveridge, theorist of the British welfare state and an earlier director of the School, was also a keen eugenicist and took the view that university teachers were self-evidently 'fit' and were to be encouraged to breed.[36]

At the same time, the Eugenics Society was developing initiatives organised around the concept of 'the promising family'. For C. P. Blacker, the secretary of the Eugenics Society, the identification of promising families offered a means of promoting positive eugenics, just as the identification of problem families (as we shall see in the next chapter) enabled the promotion of negative eugenics. For Blacker, a promising family was one in which the parents possessed 'intelligence, good physical health, moral qualities which cause them to be esteemed in their occupations and by their neighbours, philoprogenitiveness, and good family history'.[37] It was Blacker's contention that many parents who had already had 'promising' children were being prevented from having more solely because of economic pressures. He suggests therefore that various forms of assistance, including financial assistance, should be offered to 'selected parents' in order that they should provide more citizens for the future. Blacker frames his proposal as a kind of adjunct to the meritocracy, suggesting that whereas scholarships are now provided to 'help able children to make the most of their native abilities', the aim of his system would be 'to get able and well-adjusted children born' (307). His proposal is thus, in effect, for eugenic 'scholarships' for the unborn.

The Eugenics Society sponsored three surveys of 'promising families' on which C. O. Carter reported in *The Eugenics Review* in 1961. What is most striking in the report is not the final recommendation for income tax allowances and educational grants for 'promising families' but Carter's emphasis on particular qualities of character (in addition to general health and IQ) as eugenically desirable traits. He writes of 'the value of citizens who possess intelligence, energy, and also temperamental qualities which enable them to use their abilities to the advantage of their fellow men', echoing Blacker's emphasis on qualities 'esteemed by colleagues and neighbours' in his definition of promising families.[38] This emphasis on social co-operation offers a further indication of the way in which eugenic thought was being re-configured in the post-war period. Carter presents the project of differential breeding in terms of encouraging a 'genetic potential' oriented as much towards social as individual gain, thereby emphasising the convergence between eugenic interests and those of the progressive post-war political consensus. Carter also emphasises the importance of the home environment in developing this 'genetic potential'. It was this new emphasis on child rearing as well as child bearing that, it was thought, would place a particular strain on those highly educated women who would constitute the most socially valuable mothers.

The conflicts experienced by such women were the subject of Judith Hubback's pioneering study *Wives Who Went to College* (1957), a book based on the continuation of a research project begun by her mother-in-law, the educationalist and eugenicist Eva Hubback, in 1949; Carter also assisted with the research. Judith Hubback's book, which was based on questionnaires sent out to 1500 graduates, was one of the first to draw attention to the boredom of the stay-at-home mother, six years before Betty Friedan published *The Feminine Mystique*, with its identification of such boredom as 'the problem with no name'.[39] However, while acknowledging the 'frustration' felt by many graduate mothers, Hubback's study remains true to the 'sequential' model of women's dual role and (perhaps unsurprisingly in view of the connection with Eva Hubback) goes far further than Myrdal and Klein in arguing for motherhood as the primary duty of intelligent women. In a classic eugenic formulation, she argues that such women are 'very important in terms of quality. They are, moreover, a growing section of the population', and thus will contribute to a general uplifting of the race.[40] Analysing her data, she concluded with satisfaction that graduate wives had a higher fertility rate than the general population, and she was also able to establish a positive correlation between academic excellence (measured by class of degree and status of institution) and fertility.

The main thrust of her argument is that it is 'natural' for women to channel their energies into child rearing and 'healthy' for them to have as many children as possible, despite the fact that years spent bringing them up will involve an inevitable 'waste' of some of their more intellectual talents. Drawing on the language of social efficiency and evolutionary theory, she argues that:

> It is, for example, more natural to have more, rather than fewer, children. The endowment of creative energy is applied far more healthily over a larger, not a smaller, number of offspring, and the probably larger number of grandchildren. (154)

This rhetoric also informs Penelope Mortimer's *The Pumpkin Eater* (1964), a novel that explores the breakdown of an educated and intelligent mother of many children. The unnamed protagonist has brought up her children in rural surroundings, assisted by the war-time, pre-Welfare State provision of essentials for mothers and children: 'We always lived in the country, and most of the time it was the war. We ate cornflakes and eggs and carrots. . . . They went to the village school. We got free milk and orange juice'.[41] Her life has been lived according to the conviction that motherhood is inherently valuable, 'like a great tide I could be carried along on. Very strong and natural, and it seemed to me pure' (134). Motherhood is construed in terms of animal-like instinct ('She drops those babies like a cat, you know—it's a pleasure to watch') and irresistible force ('I feel myself like a torrent being damned, being forced back, turned into new channels') (37, 105). Such rhetoric is seductive but, as the novel suggests, damaging in its coercive identification of subjectivity with giving 'oneself' to others: sitting alone after her nervous collapse, Mortimer's protagonist reflects—in terms reminiscent of Woolf's Mrs Ramsay in *To the Lighthouse*—that 'I only knew how to do one thing, to give myself away. Now there was nothing left to give' (122).

Like the women studied by Judith Hubback, the protagonist of *The Pumpkin Eater* could by no means be considered a meritocrat. The 'graduate wives' of the years immediately following the war had, by definition, received their education before the 1944 Act, at a time when only middle-class families (and predominantly upper-middle-class families) would have been able and willing to pay for the higher education of girls. Nonetheless, although some of the difficulties with which these women contended were class-specific (most notably the decline in the numbers of domestic servants), others were not. The linking thread between this generation and the next was the disjunction between the constraints of motherhood and the expectations created by increasingly high levels of education. The resulting conflict was framed in similar terms by both Judith Hubback and Hannah Gavron, author of an influential sociological study *The Captive Wife*, which was published in 1966. Gavron observed that the system of education instituted by the 1944 Act attempted to offer 'equal opportunity to all children to prepare for becoming *instrumental members of a work orientated society*', but pointed out that the ideas behind this attempt 'conflict considerably with the roles and functions of motherhood as conceived by the self-same society'.[42] It would seem that this conflict was further exacerbated by the scholarship girl's experience of social mobility, which made her disinclined to step off the 'ladder of opportunity' in order to devote herself to the production of future citizens.

SCHOLARSHIP GIRLS

While the experience of the Angry Young Man and of the scholarship boy was widely represented, and became part of the dominant cultural narrative of this period, that of the scholarship girl was not, and one has to search quite hard to find her as the subject of fiction. Margaret Drabble's *Jerusalem the Golden*, published a year after *The Captive Wife*, is one of the few contemporary novels to explore the rise of a scholarship girl. Clara Maugham's intelligence (itself presented as a kind of mutation, 'some horrific deformity, like the dwarf in the circus') is the key to a university education and what she sees as 'election' to the golden city—Jerusalem/London, a world of material grace and emotional complexity. Clara's trajectory is initially cast in terms of repudiation and rejection of the 'ties of blood', of descent and origin. So fierce is her repudiation and so rapid her social progress that she begins to fear that she will crack up, inspecting herself for 'signs of manic-depression or schizophrenia'.[43] The rapid transition from the lower-middle-class world of her provincial childhood to the sophistication of the metropolis has destabilised her sense of self, creating an almost literally split personality. She survives, however, by coming to understand that her journey has not been as arbitrary as it might seem: it is not a mistake or a flash in the pan. Her trajectory has been rooted in the talents and ambitions of her mother, a pre-history that she has not recognised because her mother's hopes have been hidden and overlaid by a life of disappointments. As her mother lies on her deathbed, Clara finds some of her old exercise books that contain her poetry and stories, 'proof that she had once lived' (196). This causes Clara to feel terror at the narrowness of her own escape but at the same time a 'shocked relief' at finding her 'place of birth'. The ties of blood do not need to be repudiated: rather, they provide an endorsement of self and a support for Clara's identity. Drabble's text thus ambiguously intertwines the discourses of genetic inheritance and meritocracy as Clara, thinking of her mother, feels 'for the first time, the satisfaction of her true descent':

> Clara looked back at the writing, and at its shockingly literate echoes of stories and hymns long since forgotten, and she could bear to read no more, and she wondered whether she should fall on her knees and thank Battersby Grammar School and the Welfare State. (196)

The difficulty of shaping a narrative that reflects the gendered inflections of social mobility is also the central theme of Carolyn Steedman's memoir *Landscape for a Good Woman*, which was published in 1986. A historian who is from a working-class family, Steedman emphasises the ways in which previous accounts of post-war social mobility occluded her history and that of other working-class girls by failing to take into account the complex intersections of gender and class. Thus, she writes that the influential model of working-class family life promoted by Richard Hoggart in *The Uses of Literacy*, with 'our mam' securely at home and a masterful father meting out punishment, bore no relation to the experiences of her own childhood, in which her mother worked and was effectively a single parent while her father lacked socially endorsed power. Similarly, the major 'narratives' of psychoanalysis—for example, those of Nancy Chodorow's *The Reproduction of Mothering*—have been developed, she argues, in relation to the bourgeois family unit and cannot account for deviant patterns of mothering in more exiguous circumstances. Steedman invokes these powerful interpretive models only to undercut them, supplementing and contesting them with sections of biography, autobiography, social history and fragments of myth and fairy tale. Her aim is not to arrive at 'the' truth but to highlight the conflict between those stories that have become central and those that have been marginalised, and to make it possible for a particular 'sad and secret story' (her own and her mother's) to emerge. Steedman's choice of a largely autobiographical format in order to tell this story is also bound up with a rejection of the fictional narratives created by the 'Angry Young Men'. As she points out, women like herself, children of the 1944 Education Act and the expansion of higher education in the 1960s, simply could not see themselves in the writings of Amis, Braine or Osborne:

> We could not have talked of escape except within a literary framework that we had learned from the working-class novels of the early sixties . . . and that framework was itself ignorant of the material stepping-stones of our escape: clothes, shoes, make-up. We could not be heroines of the conventional narratives of escape.[44]

The mixed form Steedman adopts offers a more fluid discursive space, in which it is possible to improvise, to shuffle narratives and narrative forms in an effort to map the stories of those living, as Steedman puts it, 'out on the borderlands'—borderlands of gender as well as class.

Steedman's analysis of the circumstances of her childhood reveals the blind spots in Chodorow's account of the 'reproduction of mothering'. Chodorow argues that because girls are mothered by someone of the same gender with whom they identify, they develop more fluid or permeable ego boundaries than boys. Their sense of self is of a 'self-in-relationship', and it is their consequent need for close connection with others that underlies the desire to mother.[45] Steedman does not deny the existence of fluid interchanges between mother/daughter/child but points out that they are not always beneficial in their effects. So, as she explains, she does indeed identify with her mother but does not want to 'become her, and come to reproduce the circumstances of our straightened (sic) unsatisfactory life' (95). She is painfully aware of her mother's feelings of resentment over the economic and material constraints of her existence, and she is afraid that she has been shaped by and will replicate such a structure of resentment:

> My mother's sense of unfairness, her belief that she had been refused entry to her rightful place in the world, was the dominant feature of her psychology and the history she told: her life itself became a demonstration of the unfairness. (112)

Moreover, although her mother was a 'good enough' mother in her early childhood, after her sister is born, Steedman feels 'exiled from her mother's attention', given no physical affection and warmth.[46] But what is in question here is not an *individual* failure. Steedman always grounds her narrative in historical circumstance and relates the limits of her mother's capacity to love to the external pressures on her, and in particular to the ways in which she has been denied basic forms of social recognition. For example, the book opens with a memory of her mother crying in the front bedroom of a house in Streatham Hill after a health visitor has condemned the house as 'not fit for a baby' (Steedman's younger sister). This incident is just one in a series of petty humiliations that create a sense of alienation and social exclusion. Her mother accepts, though she resents, such judgements, but Steedman resolves to do 'everything and anything until the end of my days to stop anyone ever talking to me like that woman talked to my mother' (2).

She manages this, and manages to 'escape', courtesy of the Welfare State. She passes the 11 plus and goes to grammar school, and she describes the way in which a benevolent state takes over and becomes, in a very real sense, her good enough mother. She speculates that she would have been a very different person if, in the post-war years, her 'orange juice and milk and dinners at school hadn't told me, in a covert way, that I had a right to exist, was worth something'. Steedman's enthusiastic description of the state as a proxy parent is striking. Reversing the terms of the conventional opposition between biological and social parenting, she writes that if she had grown up with her parents 20 years earlier, she would have been the victim of their failures. It was the intervention of the state that gave her a sense of the worth of her own existence and 'the belief (maintained always with some difficulty) that I do have a right to the earth' (122–23). This echoes a point made in *Border Country* about the importance of having a 'social father', and in both of these texts, the sense that the state has taken over some aspects of parenting is viewed in entirely positive terms. Steedman argues that post-war welfare policies benefited her in both a material and psychological sense: she writes that 'being a child when the state was practically engaged in making children healthy and literate was a support against my own circumstances' (122). As she also points out, however, psychic structures are shaped by such interventions, and the link uncovered here, between a child's sense of self and the support of an approving state, has potentially darker implications, suggesting the potential for state intrusion into the shaping of identity.

Steedman's account vividly recalls and represents the contradictions of women's position within the meritocracy: they could 'choose' to become 'instrumental members of a work orientated society' or embrace 'the roles and functions of motherhood', but it remained extremely difficult to reconcile the two. Lorna Sage's memoir *Bad Blood* (2000) also focuses on these issues and is, like *Landscape for a Good Woman*, concerned with patterns of mothering and the (im)possibility of disrupting them. Sage tracks back from her own childhood in the 1950s to her grandmother's girlhood at the turn of the century, and uncovers a pattern of repetition through the generations whereby each of the mothers in her family has evaded motherhood in some way, shuffling the responsibility for it onto others. Through a series of substitutions involving the 'mothering' of children by fathers, husbands and grandmothers, these women have managed to hang onto the role of daughter, expecting perpetually to receive rather than to give care. As Sage puts it:

> [I]t seemed that nobody inside our family wanted to be mother, everyone was a daughter in perpetuity. You could have children of your own and still stay a mother's girl, as Grandma's example daily demonstrated. My mother, too, carried on the tradition, although she could hardly remember her grandmother, the mythic maternal figure at the start of the line.[47]

Sage's story is a remarkable one. Having started out in a village school that was run on the principle that the point of education was to 'reflect your ready-made place in the scheme of things and put you firmly back where you came from', she passes the 11 plus and goes on to grammar school.

However, she becomes pregnant at the age of 16 without even realising that she has lost her virginity. Undaunted, she revises for her A levels and takes them within days of giving birth to her daughter, eventually arriving at her version of the 'golden city', Durham University, and subsequently becoming a distinguished literary critic. Yet the price she pays, initially, is subjugation to precisely the pattern of mothering that she had wanted to reject. Like her mother and grandmother before her, Sage 'farms out' motherhood, entrusting her daughter to her mother when she goes away to study, commenting ruefully on the fact that her daughter preferred her grandparents to her 'fly-by-night' parents (275).

Writing of her pregnancy, Sage also interprets it in terms of a contemporary Denis Wheatley best-seller *To the Devil, a Daughter*, which she read 'obsessively' at this time. As the title suggests, this novel has an occult theme and revolves around a Satanist who has dedicated his daughter to the devil at birth in exchange for worldly success.[48] The novel's political subtext is more telling, however. Wheatley was passionately opposed to socialism and to post-war social change, and his fiction frequently plays on fears of the Soviet Union as a means of discrediting it by association: thus, in this novel, it transpires that the evil father has been recruited as an agent of the Bolsheviks, along with other sexual deviants. The reference to Wheatley's novel thus operates on two levels. On the one hand, it provides Sage with a pretext for a meditation on her own 'bad blood', that combination of sexual and intellectual appetites she has (allegedly) inherited from her grandfather. On the other, it also works to place this 'bad blood', circulating across generations and across genders, in the context of a new and suspect force in society, which was, just as Wheatley feared, winning through, moving up in society and entering the room at the top.

Gaining access to the room at the top was particularly difficult for women: Sage's memoir offers a dramatic example of the conflict between a career and motherhood that faced many women of her generation. Yet it would seem that the difference between the first generation of female meritocrats and that of their mothers was that, as Hilary Land has recently put it, echoing Steedman, 'our experience of the welfare state meant that we felt *entitled* to challenge the situations in which we found ourselves'.[49] In consequence, writers such as Sheila Rowbotham began to question the lack of support for working mothers, and debates were initiated over the economic costs of motherhood and childcare. As Rowbotham has subsequently suggested, these debates were extremely important in bringing women into the emerging women's liberation movement.[50] Land has also argued persuasively that there was a close and causal link between the experience of first-generation female meritocrats and the emergence of second-wave feminism, writing that:

> Paradoxically, it was the experience of being a young citizen in the early years of a postwar welfare state, based more closely than ever

> before on the principle of universality, combined with success in its meritocratic education system . . . that gave some of the generation of young women educated in the 1950s the confidence and determination to question, and to try to reform, both the 'private' world of the 'family' and the 'public' world of politics and economics. (59)

It was precisely those young women who had been successful in terms of their 'instrumental education' who brought into focus the socio-political questions surrounding 'women's two roles' that had earlier been elided.

THE RISE AND RISE OF THE MERITOCRACY

Michael Young anticipated that the meritocracy would fall in 2033, defeated by the combined forces of educated women and disgruntled manual workers. In fact, the selective education system, which he saw as the major engine of meritocracy, had been largely dismantled by the mid 1970s, partly as a result of the critique from the left with which he was associated. In 1965, Anthony Crosland, the secretary of state for education for the Labour government, issued instructions to local authorities to prepare for the introduction of comprehensive schooling, and by 1975, the 11 plus examination had been virtually abolished. However, the system of selective education did have a measurable effect on social mobility. The most recent study has shown a significant increase in 'structural' mobility, in other words, changes in the distribution of classes, during the second half of the twentieth century; Heath and Payne found a dramatic decline in the size of the working-class (from 62% of the population to 37%) and a corresponding increase in the size of what used to be called the 'white-collar class' (from 18% to 41%).[51] However, as they are at pains to emphasise, structural mobility has a positive impact on all sections of society, and it is 'exchange' mobility (mobility over and above that related to changing class distributions) that is a more significant indicator of social mobility. They found evidence of a small increase in such mobility in the birth cohorts of 1940 and 1950–59, precisely those who would have gone through the selective system of secondary education.

Obviously the meritocracy has not emerged in quite the form that Young envisaged through a highly stratified education system. However, the principle of social organisation that he outlined, whereby position and financial reward are (in theory) allocated according to merit, has become widely accepted as a model for Western democracies. A society in which wealth and social status are assigned by competition will, of course, be perceived as more equitable than one in which status is linked to inherited class position. However, as Young emphasised, such a society will continue to be structured by inequality; indeed a so-called meritocracy will legitimise inequalities that may be all the more dangerous because they are perceived to have a basis in the shortcomings of those excluded from privilege. Emphasising this point in a colloquium on Young's work, Geoff Dench has argued that the corollary of the contemporary emphasis on education as 'virtually the sole key to valued social participation' is that while government is highly responsive to the attitudes and opinions of the educated elite, it is 'contemptuous of the views and values of menial workers and the unemployed, unproductive rump of society'.[52] In a 1994 interview with Dench, Young made the same point, arguing that although equality has in some senses been extended, this is only because:

> the demands of the kinds of technology-driven society we have are for more and more highly competent people to pull levers and to design the levers. And that means an education system which is harnessed to the productive system even more closely than it was before.

While the majority of the population is infinitely better off than it was in the 1950s, those who form the 'rump of society' (the eugenicists' bottom 10%, or social residuum) remain dispossessed, and, as Young concludes, this 'has taken the heart out of millions of children as a consequence and has led to . . . some of the awful things that are happening to families and communties'. The challenge that remains is 'to take a grip on that, in the interests of a sort of equality that would be wider but not stultifying'.[53]

2 Defective Humans
Mental Deficiency in Post-war Britain

A mentally defective girl with a lick of lipstick, a flick of powder, and a cheerful fluency no matter how shallow or stereotyped, can often deceive potential employers and even magistrates into forming a considerable over-estimate of her basic intellectual powers.[1]

To categorize mentally handicapped people as 'defective' or 'subnormal' is to describe them entirely in terms of their supposed pathology, what is wrong with *them*. Such descriptions effectively mask other aspects of their social existence, or even deny them one at all.[2]

PROPHYLAXIS

'Mental deficiency' was the term used in this period to categorise those who would today be described as having learning difficulties or cognitive impairment. Up until 1959, deficiency was understood in the terms laid down by the 1913 Mental Deficiency Act, which had identified four types of defective. Idiots formed the lowest grade, consisting of those who were 'so deeply defective in mind as to be unable to guard against common physical dangers'; imbeciles were those who, whilst not idiots, were 'incapable of managing themselves or their affairs'; and the feeble-minded were those who required 'care, supervision and control for their own protection or the protection of others'. The fourth category, that of the 'moral defective', consisted of those who from an early age displayed 'mental defect' combined with 'strong vicious or criminal propensities on which punishment had little or no effect'.[3] In the 1959 Mental Health Act, mental deficiency was re-defined as 'arrested or incomplete development of mind' and was sub-divided into three categories: the severely sub-normal, those with less severe sub-normality who were 'susceptible to medical treatment' and those with a 'psychopathic disorder'. In the 1959 Act, there was a shift in emphasis away from 'care and control' and towards 'treatment and training', but definitions of mental deficiency continued to focus on the symptoms, rather than attempting to discriminate between the causes, of low intelligence and/or social incompetence. In practice this group would have included those suffering from a very wide range of clinical conditions, including brain damage caused by injury or physical illness (e.g. meningitis), Down's syndrome, phenylketonuria (inability to utilize an essential amino acid, a condition now treated by dietary measures), hydrocephaly (abnormal levels of cerebrospinal fluid) and cretinism (a form of dwarfism, now treated by the administration of thyroxine or iodine). Moreover, many of those characterised as 'mental defectives' also suffered from epilepsy or mental illness, including neuroses and psychoses, and physical disabilities were also frequently present.

Whatever classifications or indicators were used for measurement, it is clear that the incidence of 'mental deficiency' in the population was not high. For example, the statistics from the annual reports of the Ministry of Health show the total number of mental defectives 'under statutory forms of care' as 103,321 in 1947, rising to 124,246 in 1957. Statistics for the following decade (which distinguished between those receiving in- and out-patient care) showed the number of sub-normal and severely sub-normal patients in hospital beds as 61,471 in 1962, falling to 59,918 in 1970.[4] When those with milder sub-normality are included, the overall prevalence stood at around 2% of the population.[5] Yet mental deficiency caused a degree of concern among the general public, doctors, psychiatrists, educationalists and social planners that was out of all proportion to the numbers involved. This discrepancy was due to a combination of factors, perhaps the most important of which was the fact that, as we have seen in the previous chapter, eugenic beliefs about the desirability of managing the population were central to the thinking of many of the architects of post-war reconstruction. The most significant and influential figure in relation to the question of mental deficiency was C. P. Blacker, a prominent psychiatrist as well as the general secretary of the Eugenics Society from 1931 to 1952. Blacker had close links with Richard Titmuss and was instrumental in forging connections between the Eugenics Society and other post-war government advisors such as the sociologist T. H. Marshall.[6] In 1942, he was withdrawn from the army and commissioned by the Ministry of Health to write a report on the current state of the mental health services. The report was initially prompted by concerns about civilian responses to the stresses of war-time bombing, but the project expanded when it became clear that Blacker was in a position to make suggestions for the overall reorganisation of mental health services after the war. The shape and character of post-war mental health policy were thus strongly influenced by an avowed eugenicist with a particular interest in mental deficiency.

In *Neurosis and the Mental Health Services*, Blacker's recommendations were for an integrated mental health service under the aegis of the National Health Service (NHS), with increased staffing levels, improved staff training and the expansion of out-patient services. These practical suggestions were uncontroversial and became part of mainstream thinking about the provision of mental health services. In addition, however, *Neurosis and the Mental Health Services* devoted considerable time and attention to the analysis of categories of mental illness and, especially, mental deficiency.

Indeed, Blacker stretched the terms of reference of his report in order to justify this emphasis by deftly linking neurosis with 'backwardness':

> If we think of neurosis in terms of a simple and on the whole acceptable psychopathology, *i.e.* in terms of maladjustment, *it will be clear that there is no more ubiquitous cause of social maladjustment than dullness and backwardness. . . .* Every local authority has to deal with children, and every authority is aware of the problem of the dull and backward as well as of the difficult child.[7]

He goes on to argue for the 'prophylactic handling' of dullness and deficiency. Blacker's discussion of deficiency in this report was influential: for example, the prominent psychiatrist David Stafford-Clark, co-author of the textbook *Psychiatry for Students* and the best-selling Pelican paperback *Psychiatry Today*, followed closely Blacker's understanding of the causes and effects of 'deficiency' and 'unsoundness of mind'.[8] Tactically aligning himself with the agenda of the post-war Labour government, with its focus on the needs of children, Blacker stresses in his report the need for early identification of the severely sub-normal who will require specialised medical treatment; he also emphasises the need to identify 'dull' children when they are young in order to 'secure their disposal in special schools'. He envisages testing of all children at the ages of 7, 11 and 16, with a view to, as he puts it, 'screening from the bulk' those of sub-normal intelligence (52). Blacker does not go into detail about the types of care such children would receive, for as his later chapter on the 'social problem group' makes clear, his main interest is not in treatment or care but in reducing the incidence of mental deficiency within the population under the rubric of 'preventive psychiatry'.

Such an emphasis can be understood in the light of Foucault's concept of 'biological racism', as discussed in the introduction. Foucault's contention is that biological racism became inscribed within the mechanisms of the nation state at the point when the theme of the 'struggle between races' was supplanted in the nineteenth century by a 'postevolutionist' theme of the struggle for existence. A society that had previously been understood to consist of competing races was reconceived as 'biologically monist' but threatened by heterogeneous elements within; hence, the perception/fantasy of a link between the elimination of degenerates and 'deviants' and the *strengthening* of the lives of those who remained within the body politic.[9] However, as Foucault is at pains to point out, when he speaks here of 'elimination' or 'killing', he means not simply 'murder as such, but also every form of indirect murder: the fact of exposing someone to death, increasing the risk of death for some people, or, quite simply, political death, expulsion, rejection and so on' (256). It is in this sense that the treatment of the mentally defective can be viewed as a form of biological racism. The health of those segregated in mental deficiency hospitals was indeed placed at risk, due to serious overcrowding, poor hygiene and inadequate medical care. The mentally defective were also excluded from citizenship in a number of important respects. As Mathew Thomson has argued, of particular significance was the fact that they were unable to contribute personally to the funding of the newly established social insurance system. As non-participants in the social contract, the mentally defective were increasingly sidelined and consigned to what he terms 'a separate residual system for the socially and morally incompetent'.[10]

Drawing on Foucault's analysis of eugenics as a biopolitical strategy, the social theorist Nikolas Rose has suggested that in many post-war democracies, most notably Sweden,

> a kind of pastoral eugenics was practised, within the context of a developing paternalistic welfare state. It was modelled on the responsibilities of the good shepherd—who must, of course, be prepared to take sometimes harsh decisions in order to reduce the burden that weak or sickly sheep would otherwise place upon the flock.[11]

Borrowing the image of the pastorate from Foucault, Rose goes on to make the crucial point that eugenic strategies operated in these contexts in terms of persuasion as well as coercion. In consequence, throughout the twentieth century, there has been a considerable degree of overlap and blurring of the boundaries between preventive medicine and negative eugenics. This point is of particular importance for an understanding of the eugenic effects of the 'management' of the mentally defective. On the one hand, simply by virtue of being allocated to mental deficiency hospitals, those who were classified as severely sub-normal were, effectively, sterilised. There was a strict policy of sexual segregation of mental deficiency patients; moreover, many such patients spent decades in the hospital with very little opportunity for contact of any kind with the wider community.[12] These measures can be understood as a coercive form of sterilisation of the severely sub-normal on the grounds of assumed biological defect.

Equally however in the post-war period, a great deal of energy was devoted to the strategies of eugenic persuasion identified by Rose. In relation to the targeting of mild cases of mental deficiency, the mobilisation of the category of the 'problem family' was particularly significant. The term first emerged in a 1943 report produced by the Women's Group on Public Welfare on the effects of evacuation during the war: the way in which such families are described in the report points to the overlap between the 'problem family' and the so-called 'social problem group', which had been a major concern of eugenicists before the war.[13] In each case, what is being identified is a sub-set within the 'bottom 10%' of society (i.e. the bottom 10% in economic terms). This obdurate sub-set of the population is described as 'always on the edge of pauperism and crime, riddled with physical and mental defects, in and out of the courts for child neglect, a menace to the community', and is understood to be a genetically distinct group that poses a particular threat to society because of its abnormally high fertility.[14] Thus, although Blacker claimed that the shift in terminology from the 'social problem group' to the 'problem family' represented a genuine shift in understanding (with the problem family now sympathetically understood in terms of 'maladjustment'), the cause of such maladjustment continued to be presented in resolutely biological terms. The importance of these families lay precisely in the fact that they were not excluded from the community but were active—indeed over-active—participants in it, as the following description from Blacker suggests:

> The parents are often of subnormal mentality: the father may be a ne'er-do-well, the mother a conspicuously incompetent housewife . . . there is often present in either or both parents a temperamental instability which expresses itself in fecklessness, irresponsibility, improvidence and indiscipline in the home. The same quality, if evinced by the father, may produce occupational instability or long-term unemployment, alcoholic intemperance, gambling habits, and recidivism. . . . Illegitimacy and promiscuity are common. (16)

The sociopathic tendencies of such families have a damaging impact on the wider community, parental instability leading to undisciplined behaviour in the children (what might today be called anti-social behaviour) and, among the men, alcoholism and repeated criminal offences.

A related danger associated with the presence of problem families in the wider community was seen to be that of cross-breeding between a defective and a 'normal' individual. Fears about such 'miscegenation' were expressed not only by Blacker and in textbooks on mental deficiency, but circulated more widely, surfacing for example in Shelagh Delaney's play *A Taste of Honey*. Although this play is most often remembered for its exploration of an inter-racial love affair, in it the theme of mental deficiency is closely intertwined with that of race. It is introduced in the following exchange in which the main character, Jo, herself the product of an adulterous affair and now pregnant with a mixed race child, asks her mother about her father:

> *JO:* Well, was he so horrible that you can't even tell me about him?
> *HELEN:* He wasn't horrible. He was just a bit stupid, you know. *Not very bright.*
> *JO:* Be serious, Helen.
> *HELEN:* I am serious.
> *JO:* Are you trying to tell me he was an idiot?
> *HELEN:* He wasn't an idiot, he was just a bit—retarded.[15]

The shifting and evasive terminology signals the fear and unease surrounding mental deficiency: Jo also fears that 'it' (i.e. deficiency) is hereditary, and she spends the next few months wondering if her baby will inhabit 'the twilight land of the daft', as her father has allegedly done.

Through the representation of Jo's father, Delaney also plays with the idea of the link between promiscuity and deficiency. This issue is taken up in Peter Nichols' play *A Day in the Death of Joe Egg*, which activates and exposes a number of myths about mental deficiency. Joe Egg is, as her name suggests, profoundly handicapped—she is thus considered embryonic, less than human. Her father Brian fantasises about murdering her and invokes the derogatory image of the defective as a 'vegetable', describing Joe as 'a living parsnip'. Another character recommends euthanasia for Joe on eugenic grounds, unblushingly making the connection between such measures and the utilitarian 'medical' logic of the death camps:

> [I]f I say gas-chamber that makes it sound horrid—but I do mean put to sleep. When Freddie gets all mealy-mouthed about it, I say, look, darling, if one of our kids was dying and they had a cure and you knew it had been discovered in the Nazi laboratories, would you refuse to let them use it?[16]

However, it is Joe's mother Sheila who invokes the link between deficiency and promiscuity and highlights its gendered component. The link between female promiscuity and deficiency rested not so much on the perception that defective women were easy sexual prey as on a deeply ingrained folk tradition of blaming mothers for the birth of 'abnormal' children.[17] Sheila falls into this trap, attributing Joe's slow birth and subsequent brain damage to the sub-conscious guilt she experienced over having had previous sexual relationships:

> I think it was partly because I'd been promiscuous, yes, and my subconscious was making me shrink or withdraw from motherhood, all right! (28)

The eugenicists' position was that the 'daft' and the 'defective' should be strongly encouraged to control their fertility. The feminist and eugenicist Eva Hubback, in her best-selling paperback *The Population of Britain*, argued for the segregation and 'voluntary' sterilisation of mental defectives, and Blacker continued to advocate such targeting of mental deficiency patients.[18] Increasingly, however, the eugenics movement directed its attention to the expansion of preventive counselling services via the growing number of agencies dealing with child welfare, marriage guidance and family planning: in other words, they moved in the direction of persuasive rather than coercive eugenic strategies. The link between eugenic thought and the marriage guidance movement was particularly important in this respect. The Marriage Guidance Council had developed out of a sub-committee of the British Social Hygiene Council, which was in turn closely associated with the Eugenics Society.

It was first established as a national organisation in 1946, in a time of concern over rising divorce rates and an apparently falling population. Backed by members of the medical profession and the churches (with the exception of the Catholic Church, which was opposed to contraception), the Council stressed the importance of family life as a means of creating healthy future citizens. Such an emphasis on the family as a privileged site for social renewal dovetailed neatly with eugenic preoccupations and led to an alliance between the two movements that lasted throughout the 1950s. Thus, the Marriage Guidance Council actively encouraged 'pre-marital counselling' and believed that it was within their remit to provide advice for 'people with bad family records [who] will want to know whether on eugenic grounds they are justified in marrying'.[19] The Council also included within its 'general principles' the following crypto-eugenic claim:

> That parenthood normally brings to marriage, not only the fulfilment of its *racial end*, but also the achievement of one of its deepest satisfactions; and that everything possible should therefore be done to promote fertile unions. (153)

For their part, eugenicists were keen to exploit opportunities for the dissemination of their ideas via the marriage guidance movement. In 1950, for example, Margery Spring Rice (the author of *Working-class Wives* who was, like Eva Hubback, a feminist and eugenicist) approached Blacker over the question of marriage advice aimed at 'young people of subnormal mentality and physique' with the aim of preventing them from becoming 'problem families'. Blacker was, unsurprisingly, supportive of her suggestion that advice be made available through marriage guidance centres, and in 1954, a pamphlet 'Eugenic Guidance' was produced for the North Kensington Marriage Welfare Centre. This pamphlet focused on the anxieties of parents who either had 'some mental or physical weakness' that they feared would be transmitted to their children or who had already had one physically or mentally handicapped child and feared that later children might be similarly affected. The Centre offered 'expert' advice through direct contact with the Eugenics Society or with a hospital clinic recommended by them, claiming that:

> Doctors experienced in human inheritance can . . . give such parents some idea of the probability of their children being born handicapped. Often the risk will be small; but occasionally the risk may be considerable and the parents would be wise to decide to have no further children. The Centre will refer couples with problems of this kind to the Eugenics Society, or through the Eugenics Society to the hospital clinic best suited to advise on their particular problem.[20]

The leaflet also makes a coded reference to the 'problem family' in its appeal to those who plan to have a large family but 'may be uncertain whether they have the hereditary constitution to justify—or the capacity to manage—so many children'. In such cases, too, 'the Eugenics Society, through the Centre, is prepared to offer advice'.

Although this leaflet cannot be considered representative of the politics of marriage guidance, it gives an indication of the way in which the eugenics movement worked to maximise its influence in the post-war era. It continued its work of advocacy through its association with a number of organisations concerned with family welfare, including the Marriage Guidance Council, the Family Planning Association and the Moral Welfare Association. Eugenicists were also closely involved in the development of genetic advisory clinics within the NHS: the first of these clinics was opened at the Great Ormond Street Hospital for Sick Children in 1946. Throughout the 1940s and 1950s, genetic counselling was largely confined to giving the kind of advice described in the marriage welfare pamphlet mentioned above (i.e. giving parents who already had a child with a genetic disorder information about the risks associated with having another child). It was not until the late 1960s that advances in genetic science coupled with the development of amniocentesis made possible the pre-natal diagnosis of disorders such as Down's syndrome. In Britain, it then became possible under the terms of the 1967 Abortion Act to terminate a pregnancy if there were a substantial risk of the child's being seriously handicapped, mentally or physically. The public debates surrounding these developments were of crucial importance in drawing attention to the ethical dilemmas underlying all forms of 'persuasive' eugenics.

MENTAL DEFICIENCY AND SOCIAL CLASS

There is today a fuller appreciation of the complexity of the factors that affect cognitive development. In the post-war period, these factors were less well understood, and mental deficiency functioned as a relatively loose category that crossed the borders of psychiatric medicine, social medicine and education. In the discourses developed by professionals working in these areas, there was, however, a unifying thread: the assumption that 'mental deficiency' was a working-class phenomenon. The conflation of poverty and socio-pathology is clear in this description of a problem family given by R. C. Wofinden, first published in the *Eugenics Review* in 1944 but widely used for many years afterwards by social workers and medical officers of health as a guide to identifying such families:

> Often it is a large family, some of the children being dull or feeble-minded. From their appearance they are strangers to soap and water, toothbrush and comb; the clothing is dirty and torn and the footgear absent or totally inadequate. Often they are verminous and have scabies and impetigo. . . . The mother is frequently substandard mentally. The

> home, if indeed it can be described as such, has usually the most striking characteristics. Nauseating odours assail one's nostrils on entry, and the source is usually located in some urine-sodden, faecal stained mattress in an upstairs room. . . . Furniture is of the most primitive, cooking utensils absent, facilities for sleeping hopeless.[21]

The primary marker of the problem family is poverty, which is linked in turn with squalor, ill-health and low educational attainment. The key question was whether such conditions were to be viewed as the cause or the effect of mental deficiency. Predictably, Blacker continued to argue that they were the effect of a hereditary, self-perpetuating defect. So strong was his attachment to this view that, as Gillian Swanson has demonstrated, he was unable to confront or think through the evidence to the contrary provided by his colleague Hilda Lewis, who was also active in the Eugenics Society.[22] Lewis was the psychiatrist attached to an experiment in the investigation and placement of deprived children funded by the Nuffield Foundation (the 'Mersham experiment'). Although Lewis thought the parents of problem families 'shiftless, dull, psychopathic, discordant', her study concluded that their children, though of lower intelligence than the other children in the study, were freer of delinquency and neurosis and actually showed *more* social responsibility:

> An impressive demonstration of the normality of many children from 'problem families' was the speed with which, after years of drift and dirt, they fell in with the more civilised way of life at the Centre. Some of the older children were surprisingly mature and responsible . . . caring for their younger brothers and sisters and sometimes actually stepping into a protective, succouring role towards their feckless parents.[23]

Lewis links this to the 'affectionate' parenting these children had themselves received: despite the deprivation in which they lived, the affection provided them with a basis for effective social adaptation.[24] The Mersham experiment thus gave the lie to eugenic notions of hereditary incompetence, demonstrating that the children of 'problem families' could be fully integrated into 'normal' society if they were educated and supported. More importantly still, Lewis's work drew attention to the ways in which poverty could be viewed as a potential cause, not of mental deficiency itself, but of a *diagnosis* of mental deficiency. As some psychiatrists were beginning to argue, borderline deficiency, in particular, was in danger of becoming (or perhaps had long been) a proxy for perceived forms of working-class sociopathy.

This was an issue taken up by a number of reforming psychiatrists in the 1950s and 1960s, most notably L. T. Hilliard and Brian Kirman, co-editors of the textbook *Mental Deficiency* (1957, 1965), which was the first book seriously to challenge the influential account of the condition offered by Tredgold's *A Textbook of Mental Deficiency*, first published in 1908 by A.

F. Tredgold, and by 1952 in its eighth edition with his son R. F. Tredgold as author. Despite the fact that it was grounded in the medical perspectives of the early part of the century and offered case histories dating back to the 1920s, Tredgold's *Textbook* remained the standard text on the subject until the publication of Hilliard and Kirman's book, which brought together the results of more recent research into the causes and treatment of mental deficiency. For Tredgold, mental deficiency constituted a 'psychopathic diathesis' that was 'inherited and transmitted', whereas Hilliard and Kirman's contributors argued that most forms of mental defect were *not* due to hereditary factors. They emphasise the wide range of conditions that were being grouped together under the rubric of mental deficiency and argue that while those suffering from serious impairment might need continual care in an institutional context, many of those who had 'limited intellectual potential' had *no* special needs. In his contribution to the book, Kirman goes on to emphasise the shifting and permeable nature of the borderline between 'normality' and 'deficiency', and he teases out some of the social factors implicated in the diagnosis of mental deficiency. He points out that:

> a child of border-line intelligence is much more likely to be classified as mentally defective if his home circumstances are unfavourable, if he is illegitimate, or one of his parents is dead, if either of them have any serious psychiatric disturbance, or if a parent is in prison or absent for any other reason.[25]

While more conservative psychiatrists would interpret such family circumstances in terms of linked pathologies (or as Tredgold would have it 'psychopathic diathesis'), Kirman argues that there is no *necessary* link between psychiatric illness, or criminality, and mental deficiency. Rather, these conditions are likely to co-exist as a result of the external pressures of poverty and social deprivation. He also argues that a high proportion of those currently in NHS beds and classified as 'sub-normal' or 'severely sub-normal' are in fact within the normal range of intelligence, suggesting that they have been wrongly diagnosed and wrongly institutionalised. He quotes from a recent study of a sample of admissions to a mental deficiency hospital: among these patients, the mean IQ of the 'subnormal' was 71 (i.e. it was within the normal range), and among those categorised as 'severely sub-normal', at least half had IQs that should have enabled them to live independently.

The inter-relation of social factors and the assessment of deficiency is also emphasised by Hilliard in his discussion of the diagnosis of mental deficiency in women. In 1954, he conducted an experiment whereby a hundred female patients who had spent an average of 20 years in mental institutions were released with the aim of supporting themselves in the community. The experiment was extraordinarily successful, with 73 of the patients being completely discharged as a result. As Hilliard argued in an article on the study in the *British Medical Journal*, the initial diagnosis of mental deficiency in these patients was questionable as many came from large families and had had virtually no access to education: he went further to suggest that in many cases, the diagnosis of mental deficiency 'created the disease' as it led to stultifying years spent in an institution.[26] Hilliard was also able to challenge one of the most powerful myths associated with female defectives, that of their alleged promiscuity, as of the hundred 'defectives' who were released, only one became pregnant during the course of the experiment. Disentangling some of the complexities of cause and effect, Hilliard showed that the problem was not that defectives were promiscuous, but that so-called promiscuity (i.e. illegitimate pregnancy) was often in itself the trigger for a diagnosis of mental deficiency. As he put it, such women 'would pass for normal members of the community. Indeed, many of them would have continued to be considered as such had it not been for the social problems caused by the illegitimate pregnancy. They are often no less intelligent than many of their more fortunate sisters who contract a legal marriage'.[27]

In Britain, the assumption of a link between mental deficiency and social class dates back at least to the late-nineteenth-century concern with degeneration and the hereditary defects that were thought to be the *causes* of pauperism. That this association should gain renewed currency in the period after the Second World War is perhaps unexpected, but as we have seen, the links between eugenic thought and post-war reconstruction were uncomfortably close. In relation to mental deficiency, for example, a number of prominent eugenicists were closely involved with the Royal Commission on Population, which in 1949 expressed concern about an apparently inverse relationship between intelligence and fertility.[28] However, an analysis of cultural representations of mental deficiency sheds light on additional social pressures that may have helped to perpetuate this myth, in particular anxieties about social mobility and the increased political power of the working-classes. Mental deficiency figured prominently in two forms of popular fiction in this period, science fiction and detective fiction, and it is in the latter that the issue of class and deficiency is taken up—unsurprisingly, given that the policing of class boundaries had been a central concern of this genre almost from its inception.[29] In this respect, the figure of the defective, or 'sub', is put to work in complex ways in the post-war fiction of both Agatha Christie and Margery Allingham. As Alison Light has argued, Agatha Christie is a less conservative novelist than is usually supposed and certainly less conservative than 'heritage' film and TV adaptations of her work would suggest. So in her 1962 novel *The Mirror Crack'd from Side to Side*, Christie's Miss Marple is, as Light suggests, broadly aligned with a pragmatic acceptance of post-war social change and of the improved material circumstances of the working-classes (the ex-servant classes).[30] Yet despite her acceptance of the movement from deprivation to affluence, represented in this text by the brash new housing development that has sprung up around the village of St Mary Mead, Marple still condones the murder of a woman from 'the Development' on the grounds that she has unwittingly infected a celebrated actress with rubella, causing her to give birth to a mentally defective child. The text interweaves a number of themes around this central plot device. Firstly, there is the acknowledgement, common in the 'country house' novels of the period, of the decline of the hereditary aristocracy: the big house of this village, formerly inhabited by a friend of Marple's (and the setting of an earlier novel *The Body in the Library*), is now the home of the actress Marina Gregg, the representative of an emergent 'aristocracy of talent'. Secondly, the fact that Gregg's pregnancy was tainted by contact with a woman from 'the Development' suggests, of course, that the newly confident working-class might itself constitute a kind of infection, which could threaten the health and fitness of the body politic. Thirdly, the novel seems to pick up on the point made by Michael Young in *The Rise of the Meritocracy* that intelligent or talented individuals will not necessarily have intelligent or talented children. The novel's title and the epigraph from 'The Lady of Shalott' suggest that this discontinuity creates a crisis of identity for Gregg: the child fails to mirror her and shore up her sense of 'self', hence her nervous breakdown and the subsequent incarceration of the child in an institution.

Whereas Christie's text exhibits some degree of openness in its exploration of the fears surrounding mental deficiency, Margery Allingham's *The China Governess* (1963) offers a more conservative exploration of the relationship between mental deficiency and social class. This novel's engagement with the theme of post-war reconstruction is announced in its opening description of a sleek new tower block: as in *The Mirror Crack'd*, brash new buildings signal the threatening power of a more affluent working-class. The block has been built to replace a notorious East End slum, 'The Turk Street Mile', and this name, coupled with references to older buildings 'on oriental lines', foregrounds the links between the old East End and immigrant communities. Filth, squalor and animal-like regression are in turn associated with the old inhabitants: 'People lived in caves. There's no other word for them. Have you ever seen a beam eaten to a sponge by beetles? Magnify it and dress the beetles in a rag or two and that's about the picture'.[31] The gleaming new block has been built by a local councillor, Cornish, who has a personal motive for his passionate belief in the efficacy of environmental change or, as he puts it, the belief that 'decent conditions make a decent community' (16). He has a son, Barry, who was born in Turk Street and has turned out to be a mentally defective criminal. Cornish is convinced that his condition can be accounted for entirely by his deprived background.

However, Cornish's belief in the environmental causes of deficiency is undermined when the plot unfolds to reveal that Barry is not in fact his son, but the illegitimate child of a woman who is herself mentally defective. She is represented in casually damning terms as 'a bedraggled sprite of a woman with a slack mouth and huge vacant eyes, who yet managed to convey a hint of cunning' (201). In a contemporary re-working of the cradle swap plot, she has engineered a switch between her son and Cornish's in the chaos of wartime evacuation. In consequence Cornish has for years conscientiously taken responsibility for the 'difficult' Barry, unaware of the existence and fate of his biological son Tim, a likeable and intelligent young man who has been seamlessly absorbed into a wealthy adoptive family. The text thus provides two adoption case histories through which the question of the influence of heredity versus the environment (or 'nature' and 'nurture') can be tested. In one case, that of Cornish's biological son, the text suggests that he has benefited from a combination of natural wit and a privileged environment and education. To this extent he can be viewed as a representative of the meritocracy, bringing fresh energy and talent into a fading elite. Such social mobility is a phenomenon of which Allingham seems wholeheartedly to approve: Tim is rewarded in that he gets the girl (a beautiful heiress educated, like him, at Oxford) *and* the name and inheritance of the family that has adopted him. By contrast, Barry Cornish fails to respond to any of the efforts of his adoptive father and, fulfilling his hereditary genetic destiny, becomes a criminal who is in the end responsible for the only death in the text. His sub- or ab-normality is described in terms that allude negatively to the physical features of Down's syndrome and also suggest atavistic qualities:

> The face which emerged was not reassuring. It was blunt and grey, the nose springing thick and flat from high on the frontal bone of the forehead, whilst his eyes were narrow slits of dark in a tight bandage of tissue. He was not a mongol but there was deficiency of a sort there. (224)

Barry is a child from a 'problem family' who suffers from the defining characteristics of this group: 'mental backwardness, temperamental instability, and ineducability'.[32] His character is presented in terms of a regressive sociopathy and associated with ineradicable filth and defilement: 'He appeared deeply and evenly dirty, his entire surface covered with that dull iridescence which old black cloth lying about in city gutters alone appears to achieve' (224). What is particularly striking in this text is the way in which both conservatives *and* social reformers construct this problem group in terms that suggest extreme physical abhorrence and revulsion. Even the reforming Councillor Cornish, for example, describes the old Turk Street slum in these terms:

> Children crawled over each other like little grey worms in the gutters. . . . The only red things about them were their buttocks and they were raw. Their faces looked as if snails had slimed on them and their mothers were like great sick beasts whose byres had never been cleared. (107)

The powerful images of viscosity and broken bodily boundaries recall Julia Kristeva's descriptions of abjection ('I expel *myself*, I spit *myself* out. . . . During that course in which "I" become, I give birth to myself amid the violence of sobs, of vomit'[33]). In developing her account of abjection, Kristeva drew heavily on the work of the anthropologist Mary Douglas, whose 1966 study *Purity and Danger* explored the ways in which 'anomalous' conditions and 'marginal states' are managed through ritual containment.[34] It was Douglas's account of the cultural response to 'anomalous' individuals, together with Erving Goffman's analysis in *Asylums* (1961) of the stigmatisation of 'abnormal' people, which formed the main platform for the critique of the treatment of the mentally defective that began to emerge in the late 1960s.[35] Thus in 1969, in a foreword to Pauline Morris's groundbreaking study *Put Away*, the sociologist Peter Townsend described the isolation and incarceration of the mental defective in terms that closely echo Douglas's work on ritual containment: 'a potential danger to the community is removed, and a sick person is symbolically identified as different, going through all kinds of "stripping" or "degradation" procedures'. Townsend goes on to explain the public's reluctance to confront the similarities between themselves and the defective in terms that also anticipate Kristeva's work on abjection, pointing out that if people were to admit that abnormality is just one end of a continuum, they might have to accept that a sick or handicapped person is essentially no different from themselves.[36]

THE SUB-NORMAL PSYCHOPATH

The diagnostic category of the psychopathic personality became influential, even fashionable, in the post-war period. In Britain the concept is usually traced back to J. C. Pritchard's 1835 description of 'moral insanity', which he characterised as a 'morbid perversion of the feelings, affections and affective powers'—which was *not*, however, accompanied by intellectual impairment. In the 1860s, Henry Maudsley re-confirmed these characteristics of the illness, but argued that because it created an affinity with 'evil gratifications', those afflicted by it constituted a 'diseased element' in society. He also argued that the illness was hereditary, and it was at this point that moral insanity began to be linked with fears of racial degeneration. However, as Aubrey Lewis points out, Maudsley and his colleagues still made no connection between moral insanity and mental deficiency: the subsequent assumption of a link between an underdeveloped *moral* sense and an underdeveloped *intellect* seems to have grown up as a matter of legislative convenience rather than medical opinion. According to Lewis, the disease-category of moral insanity was linked with that of deficiency in the 1913 Mental Health Act for social and political reasons to facilitate 'the commitment of those defective or anti-social persons who might otherwise escape containment'.[37]

After the Second World War, the precise nature of the connection between anti-social behaviour and low intellect became a matter of intense debate. Progressive psychiatrists were sceptical of the existence of any causal link between mental deficiency and psychopathy, Kirman, for example, arguing that it was possible to make a clear distinction between anti-social behaviour resulting from deficiency and that resulting from a psychopathic disorder. He argues that the mental defective behaves like a child much younger than himself (sic) and exhibits 'quantitative inadequacy in social behaviour', getting himself into trouble just as a child gets into mischief. The psychopath, however, suffers from 'a *qualitative* disorder or perversion of behaviour, often leading to lawbreaking'. Acknowledging the difficulties of classification, Kirman nonetheless stresses that '[w]hile there is some overlap between mental defect and delinquency, which is not surprising since they have some causes in common, it is a great mistake to equate or confuse the two'.[38] The 1963 edition of Tredgold's *Textbook of Mental Deficiency* makes a similar distinction, suggesting that 'many defectives are harmless, inoffensive, tractable, and well-conducted members of the community'. Conversely, others 'are not merely subnormal, they are also markedly prone to misconduct, and this is often so vicious or criminal as to cause them to be a serious social danger'.[39] Further discussion of the causes of such 'misconduct' reveals the influence of psychoanalysis on thinking about mental deficiency, as Tredgold argues that defectives may fail to perceive the wrongness of an act because of an 'inferior ability for strong super-ego formation', while in other cases their behaviour is 'consequent on a primitive instinctual life that is imperfectly modified and indifferently controlled' (251). However, the account of 'Psychopathic Disorder with Subnormality' in this edition of the *Textbook* belies this enlightened perspective, rapidly descending into a lurid description of the behaviour of such individuals:

> [M]any of them seem almost to be possessed of a devil goading them to commit crimes. This begins when they are quite young children and persists in spite of entreaty, admonition, corporal punishment, and imprisonment. Even as children some subnormal psychopaths will commit acts of diabolical cruelty, torturing animals and those younger than themselves. . . . When they grow up, children like these may commit serious crimes such as incendiarism, train-wrecking, criminal assaults upon little girls, and homicide. (254–5)

The contradictory nature of the discussion derives in part from the composite nature of the textbook, but it also reflects the uncertainties that surrounded the definition of this disease-category, which was open to wide variations of interpretation within and between the judiciary and the medical profession. Tredgold draws attention to the dangers attendant on this in a discussion of a 1951 case in which a 'high-grade mental defective' working on a daily licence pleaded guilty to stabbing a hospital attendant five times in the neck. He was sentenced to seven years' imprisonment and, according to the judge, was 'fortunate not to be standing in the dock on a charge of murder' (259). As Tredgold notes, the worrying aspect of this case lay in the fact that the defendant had been failed by both the hospital, which did not provide the 'care, supervision and control' that he needed, and the judge, who sent him to prison rather than to the State Institution for Criminal Mental Defectives, where he could have been treated and supervised.

At least three mental defectives were executed in the 1950s, the most infamous case being that of Derek Bentley, who with Christopher Craig was charged with the murder of a policeman in 1952. Craig was 16 and hence not liable to the death penalty, but Bentley, who was over 18, was found guilty and was hanged in 1953.[40] The case attracted an extraordinary degree of media attention, largely because of contemporary anxieties about juvenile delinquency. In the words of Bill Osgerby, in the 1950s:

> Within both popular opinion and academic enquiry there arose the widely held belief that the destruction of the war, the absence of fathers and the long working hours of mothers had all contributed to a breakdown in processes of socialization, 'war babies' growing up to be post-war delinquents.[41]

Such ideas found support in studies such as Leslie Wilkins' *Delinquent Generations* (1960), which claimed that children born between 1935 and 1942 were more prone to delinquency than those born in any other seven-year period; according to John Davis, between 1955 and 1961, the numbers of young men convicted of serious offences more than doubled.[42] Bentley's trial demonstrated the confusion surrounding both definitions of mental deficiency and, again, the legal status of the defective. In his exhaustive study of the case, David Yallop shows that Bentley would have fallen into the category of the 'feeble-minded' in the terms of the 1913 Act then in force, or the 'sub-normal' in terms of the later 1959 Mental Health Act. On two separate occasions he had been tested and found to have an IQ score of below 70, and he had been rejected for National Service on the grounds of his very low intelligence. In addition, he had suffered from serious attacks of epilepsy since childhood. Thus in the opinion of James Munroe, the head of psychological medicine at Guy's Hospital who examined him in 1949, he was 'feeble-minded'.[43] However, Dr Matheson, the medical officer of health for Brixton Prison who assessed Bentley's health before his trial, discounted Munroe's expert opinion. Without giving any evidence for his view, he stated that Bentley was *not* feeble-minded, and the evidence of Munroe's earlier diagnosis never reached the court. The facts quickly became public, however, and it was largely as a result of this case, and the widespread perception that Bentley had been unjustly treated, that diminished responsibility first became admissible as a defence in cases of murder in the 1957 Homicide Act.

Yallop also cites evidence to suggest that Christopher Craig had a psychopathic personality, in the terms in which the condition was then diagnosed and understood. The most common means of assessing psychopathic personality in this period was by an electroencephalograph (EEG) examination, as it was thought that psychopaths displayed an abnormal pattern of electrical activity in the brain. According to an EEG examination carried out at the Maudsley Hospital, Craig showed psychopathic characteristics consistent with abnormal development, as well as having 'a history of bed-wetting in late childhood, deep sleep and personality immaturity' (139). Craig's examination took place immediately before the trial, but again the report did not go forward to the court. This chain of events highlights—among other things—the gulf between general physicians and psychiatrists, the latter having a far greater appreciation of the complexities of diagnosis with reference to mental deficiency. No such appreciation was evident in the press coverage of Christopher Craig's case. Whereas Bentley's 'deficiency' was rarely mentioned, the press delighted in presenting Craig as the embodiment of the sub-normal psychopath. Newspaper reports emphasised his illiteracy and consumption of American horror comics, thus tying together his allegedly low intellect and his interest in violent crime. Yallop quotes too from a *Sunday Pictorial* feature that juxtaposed Craig's poor school reports with a description of his appearance, which draws on all the clichés of pre-war gangster films:

> The pale yellow walls around him gave a sinister pallor to his white face. He tossed back his coal-black hair, turned his strikingly handsome head. . . . And then he grinned and broke the spell. As I saw the left lip curl, and the impudent flicker creep into his eyes, I knew I was watching just another of those brass-faced, little hooligans. (214)

Aspects of the relationship between Bentley and Craig are replicated in the relationship between Alex and his 'droogs' in Anthony Burgess's 1961 novel *A Clockwork Orange* (filmed by Stanley Kubrick in 1971). The potential sources of Burgess's novel have been much discussed: it has been linked with his distaste for the youth culture of the 1950s and also with a traumatic incident during the war when his wife was assaulted by GI deserters and suffered a miscarriage as a result. However, the link between Burgess's text and contemporary concerns over mental deficiency and psychopathy has been overlooked. The two main characters are psychopaths: Alex is an abnormal rather than a sub-normal psychopath; in many respects his intelligence is above the average, as is demonstrated by his linguistic verve and his fondness for classical music. However, he lacks affect and a capacity for remorse, and he indulges in that 'ruthless and determined pursuit of immediate gratification', which, according to David Stafford-Clark in *Psychiatry Today*, was the principal characteristic of the psychopath. In particular, Alex's sexual and emotional immaturity lead him to indulge in what

Stafford-Clark calls 'deliberate perversions . . . wanton repeated and joyless seduction, and many of the more grotesque and outrageous sexual crimes' (117–18). His sidekick Dim is by contrast a sub-normal psychopath, and he is represented through the lens of Alex's narrative in terms which suggest that he is barely human. That Dim is 'really dim' we learn in the second sentence of the novel; he is 'beyond all shadow of a doubting thomas, the dimmest of we four' (4); he is 'gloopy and a domeless wonderboy' who 'died before he was born'.[44] The novel emphasises the intellectual gulf between Alex and the illiterate Dim in a scene in which Alex feels, as he bathetically puts it, as though 'a great bird had flown into the milkbar' as a woman sings a few bars from a fictional German opera. Dim's response is to howl like a dog and blow a raspberry—leading Alex, 'drowning in redhot blood', to re-affirm his kinship with Dim by hitting him hard in the mouth.

The novel is linked with the Bentley and Craig case in another respect. The strong public feeling generated by this case was largely due to the fact that the victim was a policeman, and press comment on it played on contemporary fears of conflict between amoral youth and the police as the representatives of social order. In *A Clockwork Orange* Burgess offers a more critical view of the forces of the law: indeed, he takes the opposition between the policeman and the psychopath and empties it out, as the vicious (sub-normal) Dim moves seamlessly, in the second part of the novel, to take on the role of policeman. The acts of violence that had characterised his behaviour as Alex's droog fit equally well with his policing activities, as we see when Alex watches him and Billy 'lashing into these starry old vecks with great bolshy glee and joy' and 'smecking with the fun they'd had' (109). The implication is that morally speaking there is little to choose between the random brutality of the young thug and the socially sanctioned (but inefficacious) brutality of the police.

Burgess's attitude to mental deficiency in the novel is ambivalent. On the one hand, it is Alex's vicious attack on Dim, prompted by Dim's signal failure to demonstrate any aesthetic sensibility, that leads to Alex's punishment (twice over, firstly when he is imprisoned and secondly when he is badly beaten up), which would seem to suggest some textual sympathy with Dim. On the other hand, as in Allingham's *The China Governess*, mental deficiency is also strongly associated with disgust and the abject. When Dim has his turn in raping Alexander's wife, Alex describes him as doing it 'in a beasty snorty howly sort of a way'. Later Dim urinates on the fire and wants to 'dung on the carpet', though he is stopped by Alex, whose physical distaste for Dim is also signalled through references to smell: when the droogs are on their way back after this evening out he notes that Dim is 'looking all dirtied over and too much von of sweat on him, which was one thing I had against old Dim' (20–1). Alex's need to 'abject' Dim becomes most acute at the point after he has been brainwashed and then taken in by a former victim of his violence, F. Alexander (a reference to the inventor of the Alexander technique for improved posture). The brainwashing has deprived Alex of that power of moral choice, which for Burgess defines humanity, leading to a potential collapsing of the boundaries between Alex and his droog. Alex is appalled to feel that he has been reduced to Dim's level and struggles to distinguish himself from him:

> And he began to stroke my left rooker as if I was like an idiot, grinning in a bezoomny way. I creeched:
>
> 'Stop treating me like a thing that's like got to be just used. I'm not an idiot you can impose on, you stupid bratchnies. Ordinary prestoopnicks are stupid, but I'm not ordinary and nor am I dim. Do you slooshy?' (121)

Precisely because the novel attacks the behaviourist psychology of B. F. Skinner and the 'social management' of the population through psychological conditioning, it has been read as a critique of all forms of 'people control', including that of eugenics.[45] However, Burgess's privileging of self-conscious ethical choice as *the* defining characteristic of the human has as its corollary the assigning of those unable to make such choices to the category of the less-than-human. In this respect it is striking that it is the reproductive future of Alex, not Dim, which is brought into play at the end of the novel. Burgess emphasises the link between Alex's redemption, which may owe something to the psychological treatment he has received in prison, and his ambition to settle down and have a son. The implication is that the intelligent and reformed Alex has a right to such a future while Dim does not.

THE HUMANIST FRAME

In addition to those working in psychiatry and/or mental health, a number of biologists took an interest in the question of mental deficiency in this period, often framing this 'problem' within a eugenic perspective. For example, when the distinguished immunologist Peter Medawar gave the BBC's 1959 Reith Lectures, he devoted one of them to the subject of 'Intelligence and Fertility' and in it concluded that the average level of intelligence was declining. He was at pains to refute the popular notion that, as he put it, 'idiots and imbeciles are riotously fertile', pointing out that the majority were either sterile or confined to institutions. However, in his view a problem had arisen 'over the greater fertility of those who are somewhat below the average of intelligence', so that 'their progeny are tending to crowd the rest of the population out'.[46] As he points out, steps could be taken to reverse this trend, and changes in taxation, family allowances and educational grants had already addressed some of the factors preventing the more intelligent from having large families. Medawar also went out of his way to distance himself from crude eugenic beliefs in 'desirable "inborn" characteristics', arguing that patterns of human inheritance were far more complex and unpredictable than eugenicists (and some geneticists) would like to suggest. Nonetheless, he emphasised throughout these lectures the importance of finding a '"humane" solution of the problems of eugenics, particularly of the problems of those who have been handicapped by one or another manifestation of the ineptitude of nature' (95).

These ideas are taken up in Julian Huxley's 1962 Galton Lecture 'Eugenics in Evolutionary Perspective', republished in his 1966 Pelican paperback *Essays of a Humanist.* Huxley had been professor of zoology at King's College London, then professor of physiology at the Royal Institution before in 1946 becoming the first director-general of UNESCO. He had also developed a high profile as a 'public intellectual' through his popular science books and appearances on television in such programmes as 'The Brains Trust'.[47] In his Galton Lecture he argued that negative eugenics had become 'increasingly urgent' in view of two new factors: firstly, mutations caused by atomic fallout, and secondly, the 'increased survival of genetically defective human beings' due to advances in medicine and social welfare. Such defective individuals included those with mental defect and low IQ, for whom Huxley proposed a number of sterilisation measures. For those with a very low IQ, 'voluntary' sterilisation was 'probably the best answer'. In the case of the defective married male, this should be accompanied by 'artificial parenthood' via donor insemination.[48] Huxley was an enthusiastic campaigner for donor insemination, or AID, because of the opportunities it offered for the 'deliberate encouragement of superior genetic endowment'; he also looked forward enthusiastically to the time when the ova as well as the sperm of the healthy and intelligent could be frozen and transplanted in the service of the improvement of the race. Where 'voluntary' sterilisation was not possible (in other words, in cases where the mentally defective were not under medical supervision), Huxley advocated the use of compulsory contraceptive vaccines (273).

In Huxley's view, the possibility of such interventions signalled nothing less than the advent of a new phase of evolution, as for the first time human beings had developed enough knowledge and understanding to take control of their own and the planet's destiny. In his introduction to *The Humanist Frame* (1961), a collection of essays he edited which included contributions from Michael Young and the geneticists H. J. Muller and C. H. Waddington, he suggests that a shift is taking place as significant as that which occurred when ancestral species first moved from being amphibian:

> Turning the eye of an evolutionary biologist on the situation, I would compare the present stage of evolving man to the geological moment, some three hundred million years ago, when our amphibian ancestors were just establishing themselves out of the world of water. They had created a bridgehead into a wholly new environment.[49]

According to Huxley, human beings are in transition between a 'biological' stage of evolutionary progress and a new, 'psychosocial' phase, which will operate by 'the mechanism of cultural tradition' (16). This argument bears some similarities to Richard Dawkins' idea of the 'meme' as a unit of cultural transmission, but what is most striking in Huxley's account is his concentration on science as the primary force through which man can become 'the sole agent for the future evolution of this planet' (17). The advances of science have, he argues, brought about the possibility of a step-change in evolution, so that '[h]uman destiny need no longer be merely an affair of hopes and fears. In principle, it can be rationally defined on the basis of scientific knowledge, and rationally pursued by the aid of scientific methods' (38). These claims can be read as early manifestations of the transhumanist movement that has become prominent over the last two decades; indeed the term transhumanism was coined by Huxley in 1957 when he defined it in this way:

> The human species can, if it wishes, transcend itself—not just sporadically, an individual here in one way, an individual there in another way, but in its entirety, as humanity. We need a name for this new belief. Perhaps *transhumanism* will serve: man remaining man, but transcending himself, by realising new possibilities of and for his human nature.[50]

Transhumanism continues to be associated with the belief that the human species can transcend itself through the exercise of rational choice. Twenty-first-century transhumanists reiterate Huxley's claim that human beings can take control of evolution and are confident that human reason—in the guise of biotechnology—will bring about a future in which humans will live 'better' lives as a result of their enhanced physical and mental capacities.[51] Critics such as Francis Fukuyama argue, however, that the movement threatens the foundations of liberal democratic society, which is grounded in 'the empirical fact of natural human equality'; significant enhancements for some would put pressure on the principles of common humanity and equal rights.[52]

Huxley and his colleagues were not alone in pinning their hopes for a better future on scientific progress, but in the wake of the Holocaust and atomic bomb, there was also considerable anxiety about the nature and purpose of scientific enquiry. It was the science fiction of the period that engaged most fully with the tension between belief in scientific progress and fear of destruction, the work of J. G. Ballard being exemplary in this respect. Ballard's first four novels fall into the category of catastrophe fiction, which as Roger Luckhurst has argued is a peculiarly British sub-genre of science fiction that dates at least as far back as H. G. Wells' *The War of the Worlds* (1898).[53] Such fictions are characterised by the rapid development of a global catastrophe, which causes cities and states to collapse and the veneer of social order and civilisation to be completely stripped away.

The novel usually begins after the catastrophe and focuses on the struggles of a few plucky survivors. *The Wind from Nowhere*, *The Drowned World* and *The Crystal World* all conform to this pattern, but it is in *The Drought* that Ballard focuses on the cultural anxieties that were converging in this period on the figure of the defective.

The catastrophe in *The Drought* has been brought about by the unchecked development of scientific industry, as for decades contaminated and radioactive waste has been discharged into the sea, causing a film to form over its surface, which has prevented cloud-formation and led to world-wide drought. This narrative trajectory is familiar in science fiction, of scientific hubris followed by 'natural' nemesis. However, the distinctive emphasis in this text is less on the damage to the earth's eco-system caused by the drought than on the possibility that 'man' himself may have to change, or mutate, as a result. Ballard thus offers a reverse image of the future envisaged by Huxley; this is a future in which man's pursuit of scientific and technological progress has led not to the enhancement of human life but to its potential destruction. Developing this theme, Ballard sets up a complex interplay between two key figures in the text: the doctor, Ransom, who is associated with enlightened rationality, and the 'idiot', Quilter, who represents unreason or at least the lack of rational capacity. In the opening paragraph of the novel, the reader is closely aligned with Ransom's perspective as he looks down on Quilter:

> He saw Quilter, the idiot son of the old woman who lived in the ramshackle barge outside the yacht basin, standing on a spur of exposed rock on the opposite bank and smiling at the dead birds floating in the water below his feet. The reflection of his swollen head swam like a deformed nimbus among the limp plumage.[54]

Quilter appears an anomalous creature, set apart from civilisation, and yet the reference to a 'deformed nimbus' suggests other possibilities, for a nimbus is a halo surrounding the head of a deity, and it also refers to a cloud, presaging rain. Together these connotations suggest that Quilter might be read not as an idiot but as an alternative god who could bring rain/redemption. Indeed, as the novel unfolds and Ransom and Quilter journey back and forth across the arid landscape, it becomes increasingly unclear who is better equipped to survive within it, the doctor who possesses the powers of instrumental reason or the 'idiot' who lacks them. As the two circle each other, the balance of power between them keeps shifting, so that at several points the perspective of the opening is entirely reversed. For example, in the third part of the novel, Ransom abandons his struggle for survival at the coast and goes back to the desert, a journey that he sees as a movement towards a future defined, paradoxically, only in terms of the past, where 'the unresolved residues of the past would appear smoothed and rounded, muffled by the detritus of time, like images in a clouded mirror' (186). In a deserted town, staring in at the windows of an abandoned shop full of blank-faced mannequins, he catches sight of Quilter, who is now watching and indeed stalking him:

> Abruptly, Ransom caught his breath. Among the blank faces, partly obscured by the reflections of the building behind him, was a grinning head. It swam into focus, like a congealing memory, and Ransom started as a shadow moved in the street behind him.
>
> 'Quilt–!' He watched the empty streets and pavements, trying to remember if all the foot-prints in the sand were his own. (188–9)

What is at stake in this encounter is this: does Quilter represent a grotesque past (as in 'a congealing memory') or does he represent the biological future? If the ambiguous relationship between Ransom and Quilter is read in terms of a post-Darwinian struggle for survival, Quilter wins the contest hands down and represents the (degenerate) biological future. The text is threaded through with allusions to *The Tempest* but departs from Shakespeare's plot when Quilter-as-Caliban succeeds in taking the sister of one of Ransom's friends—named Miranda—away from her family and seducing her. He and Miranda not only survive but thrive in the drought, literally feeding on the carcasses of dead sailors to produce 'deformed' children. These are represented in terms of deficiency and monstrosity but also as possessing, like Quilter, an uncanny power:

> Their swollen heads and puckish faces were perfect replicas of Miranda and Quilter. Each had the same brachycephalic skull, the same downward eyes and hollow cheeks. Their small necks and bodies seemed barely strong enough to carry their huge rolling heads. To Ransom they first resembled the children of the congenitally insane, but then he saw their eyes watching him. Half asleep, their pupils were full of dreams. (208–9)

The references to *The Tempest* have a double resonance, drawing both Shakespeare and *Brave New World* into the frame of the text; the title of Huxley's novel comes from *The Tempest*, and the play is a crucial point of reference throughout.[55] However, the echoes of *Brave New World* serve, paradoxically, to highlight the specificity of the dystopia presented by Ballard. In *The Drought* reproduction is not tightly controlled, with 'babies in bottles' but is, precisely, uncontrolled, and it appears by the end of the novel that some kind of unforeseen mutation is taking place as a result of the environmental pressures depicted in it.[56] Quilter's children can be seen as representing a new strain of humanity: they are 'idiots' who may lack the rational intelligence so valued by the contributors to *The Humanist Frame*, but they survive the rigours of the drought through intuition, instinct and cunning. The implication is that these children will populate the new world that emerges when the rain begins to fall at the end of the novel.

The Drought invokes a number of interpretive frameworks that it holds provocatively in suspension. As many commentators have noted, it bears the traces of Ballard's interest in psychoanalysis and surrealism, signalled through the repeated allusions to Yves Tanguy's painting *jours de Lenteur*, an image that for Ransom represents freedom from 'the tiresome repetitions of everyday life'. The novel also draws on contemporary philosophy, in particular Heidegger's insights into the relationship between being and time, and a Christian perspective is invoked via Ransom's name, which alludes to Christ's blood and the redemption of humankind. However, the framework of evolutionary biology is at least as significant. In *The Drought*, as in his other novels, Ballard gives full weight to a post-Darwinian understanding of man as a species that must adapt in order to survive and that will in the course of time be superseded. In this respect the novel challenges the anthropomorphism of Huxley and other transhumanists, whose claims for the exceptionality of human beings are in tension with Darwin's view of evolution as a process of undirected flux and change. Ransom's acceptance of this explains his 'surrender' to Quilter at the end of the novel, as he walks into the towering sand dunes 'gleaming with the bones of untold numbers of fish' and into an 'immense pall of darkness'. His abandoning of the field of struggle and his absorption into the landscape of evolutionary history also chimes with Ballard's scepticism about the power of human reason, memorably expressed in his memoir *Miracles of Life*. Reflecting on his experience of wartime Shanghai after the Japanese invasion and the effect on him of the Nuremberg trials of Nazi war criminals, he writes of the gap between his pessimism and the faith in progress he saw all around him:

> The faith in reason and rationality that dominated post-war thinking struck me as hopelessly idealistic, like the belief that the German people had been led astray by Hitler and the Nazis. I was sure that the countless atrocities in eastern Europe had taken place because the Germans involved had enjoyed the act of mass murder, just as the Japanese had enjoyed tormenting the Chinese. Reason and rationality failed to explain human behaviour.[57]

Doris Lessing's *The Four-Gated City*, published in 1969, offers a similar view of humanity as a washed-up species prone to a self-destructive violence, which is likely to be its undoing. Like Ballard, Lessing views human beings from a long-term evolutionary perspective, and she too is interested in the ways in which the human species might mutate and change. There was considerable contemporary anxiety about genetic mutation caused by atomic radiation, but Lessing converts this threat into an opportunity, envisaging a future in which atomic fall-out has stimulated a new phase of human evolution. At the end of *The Four-Gated City*, Martha Quest (the central protagonist of this novel and of the sequence of which it is a part) describes the range of new 'types' of human beings who are being born in the aftermath of the atomic attack:

> Our children, particularly those born later, grow more and more diverse. If an organism is shocked by a dose of sudden radiation (as they used to bombard atoms in the laboratories with neutrons to change their structure) then perhaps it may become ill and bleed, turn idiot, or develop in any one of a number of unknown ways?[58]

Lessing here imagines a shake-up of genetic material that distributes human characteristics across a far greater range, with more 'idiots' at one extreme and children who are 'quick and sensitive' at the other. However, the picture is complicated by the fact that even those Martha calls 'idiots' have unusual gifts; although they do not talk, they communicate with each other wordlessly. In this respect they are rather like Quilter's children in *The Drought*, who have a telepathic ability to read 'ordinary' human minds. For both writers the 'idiot' or 'defective' is an unsettlingly ambivalent figure, standing outside moral norms but possessing extra-rational powers that seem to confer a distinct advantage in the 'struggle for existence'.

CHANGING PUBLIC POLICY: BEYOND THE ASYLUM

In 1961, Enoch Powell, then minister of health, announced a dramatic shift in policy in relation to the mentally ill. He wanted to move mental patients away from large institutions, integrate mental health treatment with general health care and encourage therapy for the mentally ill, wherever possible, through 'community care'. Powell's proposals met with strong opposition from those who feared that in practice this policy would mean that mental hospitals and mental health services would be left to decline and disintegrate. However, in many respects, Powell's proposals caught the mood of the moment: this was a time of optimistic belief in the power of the new psychotropic drugs, and libertarian resistance to institutionalisation, deriving partly from Erving Goffman's work on the adverse effects of the 'total institution' and partly from the influence of the anti-psychiatry movement.[59] Powell's proposals thus gained support from those on the political left as well as the right and were accepted and gradually implemented. Increasingly, mental patients were treated in wards and day units in general hospitals, and mental hospital populations slowly declined.

When the NHS came into being in 1948 it took over responsibility for all mental illness, so that in theory Powell's proposals applied not just to large general asylums but also to long-stay mental deficiency hospitals. As Kathleen Jones has pointed out, for a while these specialist hospitals remained 'quiet backwaters' unaffected by plans to dismantle the larger asylums.[60] However, mental deficiency hospitals became a focus of public concern in

1967 when the *News of the World* published allegations of staff cruelty and the neglect of patients at the Ely Hospital in Cardiff. Similar allegations were made about Farleigh Hospital in Somerset, and the Department of Health and Social Security set up committees of enquiry. In both cases, allegations were upheld, and staff members were prosecuted. In 1968 the publication of Pauline Morris's detailed study of mental deficiency hospitals, *Put Away*, drew further attention to the difficulties and contradictions involved in the provision of institutional care for 'defectives'. Morris found that the care provided in many hospitals was inadequate. She uncovered low morale and poor communications among the nursing staff and described the physical conditions as 'barrack-like' with 'spartan provisions'. She also suggested that the majority of 'patients' in these hospitals needed social support rather than medical treatment, arguing that:

> because our culture is most ready to accord tolerance to those who are defined as being in some way ill, the subnormal person is most readily accommodated by the convenient fiction that he is, first and foremost, in need of some medical attention. Hence what may be no more than an institution for the residential containment of the subnormal is defined as a 'hospital'.[61]

Morris drew on the work of Goffman to highlight the particularly damaging effects on such patients of life in a 'total institution', where the possibilities of individual growth and development were drastically curtailed. Similar points were made in Joanna Ryan and Frank Thomas's powerful polemic *The Politics of Mental Handicap*, published in 1980, which gave a first-hand account of nursing in a mental deficiency hospital where the living conditions were filthy and the patients were often given high-powered tranquilisers to make them more docile and accepting of the institutional regime.[62]

As a result of publicity about the conditions in the hospitals and pressure from support groups such as the National Society for Mentally Handicapped Children (now Mencap), a policy of deinstitutionalisation was implemented. The main focus was on moving people with mild disabilities (the old 'feeble-minded') into an expanded network of hostels, semi-supported group homes, family placement schemes and independent living. At the same time, a number of hospital-based community units were established for those with more severe disabilities. Thus, while in 1976 there were more than 50,000 people with learning disabilities living in mental deficiency hospitals, by 2004 this number had dropped to around 4,000.

The movement of mental deficiency patients out of the hospital and into the community was accompanied by a number of changes in educational policy. The 1944 Education Act had required local authorities to educate all 'handicapped' children but excluded children with severe mental deficiency from this provision. In response, parents and volunteers set up a number of voluntary training schools, many of which eventually gained support and funding from local authorities. The 1970 Education Act placed the responsibility for educating all mentally handicapped children with local authorities and recommended that it should take place in special schools. In 1978 the Warnock Report on Special Education recommended inclusion in mainstream schools wherever possible, and since that time debates have continued over whether children with learning difficulties should be taught in mainstream schools or in separate institutions.[63] In a pamphlet published in 2005, for example, Warnock argued that despite the good intentions behind the inclusion of most children in mainstream schools, one of the unintended consequences was that special schools had become mere 'places of containment' for the severely disabled.[64] Overall, however, it seems that the move towards educational integration, together with deinstitutionalisation, has led to better outcomes, in both psychological and material terms, for those with what are now called learning difficulties.[65] As the change in terminology suggests, 'mental deficiency' is no longer defined in pathological terms and in consequence can no longer be represented as a hereditary 'disease' that threatens the health of the body politic.

3 Genetics and Eugenics

> The 'gene' is nothing but a very applicable little word, easily combined with others, and hence it may be useful as an expression for the 'unit factors,' 'elements' or 'allelomorphs' in the gametes, demonstrated by modern Mendelian researches.[1]

> There are those who will feel that the idea of genetic determinism is yet another attempt to comprehend the complexity of life under one simple formula. Indeed the formula is simple. But, in applying it, we certainly make no pretence to simplicity. It is merely, we dare to think, not beyond the wit of man to grasp.[2]

> It is surely clear that good health, high intelligence, general benevolence . . . are desirable qualities which we would all agree on. We would agree also that these qualities are not uniformly distributed. There are people who are deficient in intelligence, for example (I mention intelligence because this is something we can to some extent measure). Surely it is a very reasonable aim for us to try to increase that.[3]

It has been argued by a number of historians that genetic science successfully dissociated itself from eugenics in the post-war period. Daniel Kevles, for example, writes that 'the revelations of the Holocaust had all but buried the eugenic ideal' and that under Lionel Penrose's leadership, the staff of the Galton Laboratory 'bristled with contempt for the country's remaining eugenic activists'.[4] Similarly, Richard Soloway argues that the 'eugenic consensus' that had pervaded British scientific thought for decades broke down in consequence of the revelation of Nazi atrocities.[5] Yet this argument does not stand up to closer scrutiny. From the outset the eugenics movement provided an important enabling context for the development of genetic science. Galton himself conceived of eugenics as a science rather than as a social or cultural movement, and his statistical work on human heredity paved the way for Karl Pearson's research in biometrics, that is, the statistical study of evolution and heredity. Galton encouraged Pearson's research and set up the Galton Laboratory in 1905 to take the new field forwards. Biometrics was soon challenged by the re-discovery of Mendel's work, which focused on the segregation of 'unit characters' or genes and laid the foundations for what became known as the Modern Synthesis of evolution and genetics. Under Pearson, the Galton Laboratory resisted Mendelianism, but when R. A. Fisher succeeded Pearson as Galton professor he brought mathematical and Mendelian genetics together, and under his leadership the Laboratory became world famous for its research in human genetics. Throughout this period the eugenics movement and genetic science were thus closely intertwined. Their aims overlapped, for both were concerned with the study of heredity with the ultimate aim of intervening in the process. Many key geneticists were also involved with the Eugenics Society, including Fisher, Julian Huxley and J. B. S. Haldane.

When Penrose succeeded Fisher after the war he made efforts to free genetics from its association with eugenics, being acutely aware of the stigma of the connection. He had his chair renamed the Galton Professorship of Human Genetics (rather than Eugenics) and changed the name of the Laboratory's journal from *Annals of Eugenics* to *Annals of Human Genetics*.[6] Yet it remains hard to separate the conceptual framework of post-war genetics from eugenic assumptions and beliefs. As noted above, the scientists who were researching the molecular structure of the gene aimed to uncover the mechanisms of heredity, a goal that had been central to the eugenics movement from its inception. Some of those working in human genetics were pursuing such knowledge for its own sake, while some, like Penrose, were investigating potential therapeutic interventions. But many were interested in the other side of the coin, that is, the potential of genetics to improve the quality of human life. One of these was J. B. S. Haldane, who combined left-wing politics with a sustained commitment to both positive and negative eugenics; another was Julian Huxley, whose research in evolutionary biology led to an influential second career as a government advisor and public intellectual. Huxley continued to endorse the core beliefs of eugenics and to argue that genetic science had a responsibility to improve the future of humanity. As we have seen in the previous chapter, he tirelessly promoted the idea that advances in the biological sciences would soon make it possible for human beings to shape their own nature.[7] Such views were by no means confined to the older generation of geneticists. In 1963, for example, Francis Crick participated in a conference organised by the CIBA Foundation on 'Man and His Future'; as the discussant in a panel on dysgenic fertility, he argued that genetic deterioration posed a significant threat to the human race. One solution he proposed was a tax on children, which would deter the poor from reproducing, the assumption being that the rich were better endowed with desirable qualities of health and intelligence. Alternatively, he suggested that there could be a licensing system for reproduction and that those who were 'genetically unfavourable' should be allowed to have only one child.[8] James Watson's eugenic commitments have become widely known through his disparaging remarks on race in 2007 but date back over decades.[9]

The high public profile of figures like Huxley and Crick challenges C. P. Snow's contention in his 1959 lecture 'The Two Cultures' that this was a period in which British culture remained backwards-looking and resolutely anti-scientific. Yet as Stefan Collini has pointed out, Snow's interpretation of the relationship between the arts and the sciences was skewed by his life experiences and class position.[10] As a scientist from a lower-middle-class background, Snow felt patronised by what he saw as the traditional humanities elite, and this seems to have led him to overestimate its cultural dominance. Conversely, his success in science led him to exaggerate the open, meritocratic nature of the scientific enterprise, at least in relation to the biological sciences. As Daniel Kevles points out, the success of post-war genetics in Britain owed a great deal to the strong personal community that grew up among the leading figures. Strikingly, most members of this community were members of a hereditary elite, connected through shared education and family ties—a fact that did not escape the young James Watson, who noted that in England, 'all the important people, if not related by marriage, seemed to know one another'.[11] In this respect, the genetics network can be seen as something of a scientific equivalent of the Bloomsbury Group. Like the 'Bloomsberries', the main players had cultural capital and powerful connections, which they exploited in order to make the case for the centrality of biological science to the re-shaping of post-war society. For example, both Huxley and Haldane were regular radio and TV broadcasters, reaching large audiences through such programmes as 'The Brains Trust', and both contributed regularly to newspapers and periodicals. Huxley, Haldane, Darlington and the embryologist C. H. Waddington wrote best-selling Pelican paperbacks, introducing genetic ideas and their thinking about genetics and society to the wider public. These books were lucidly written and widely read: Waddington's *The Scientific Attitude* sold 100,000 copies in 1941 alone, Darlington's *Genetics and Man* (1964) and Huxley's *Essays of a Humanist* (1963) went through several re-printings, while Watson's *The Double Helix* (1968) was an international bestseller. The influence of popular science writing has been largely overlooked in studies of post-war literature and culture, but, as noted in the introduction, it was an extremely significant factor in the dissemination of scientific thought, facilitating what we would now call the public understanding of science. Doris Lessing, for example, recalls being 'thrilled' by Haldane's pieces for the *Daily Worker*, 'explaining new discoveries in science. . . . I knew people who bought the paper for these articles and read not a word of the rest'.[12]

This chapter explores the politicisation of genetic science during the Cold War and examines the close connections, on both sides of the ideological divide, between theories of biological and social organisation. It draws out the links between the persistence of eugenic ideas and assumptions in the West and the predominance of a gene-centric model of inheritance, and it contrasts this with the more Lamarckian views of inheritance that prevailed in the USSR. It then explores responses in science fiction texts to the scenarios of genetic development and change outlined in popular science writing of the day.

COLD WAR GENETICS: THE LYSENKO AFFAIR

The importance in post-war society of science in general, and genetics in particular, is clearly demonstrated by the controversy that blew up in 1948 around the research of Trofim Lysenko, a geneticist working in the Soviet Union. This had far-reaching consequences, and in one of Foucault's early discussions of biopower, he notes that this 'sordid affair' first sparked his interest in the relationship between power and scientific knowledge.[13] Foucault reads Lysenko as an example of a new kind of intellectual, the 'specific intellectual' or 'absolute savant' produced by the rapid development of science in the twentieth century and the 'technico-scientific structures' associated with it. Unlike the 'universal intellectual', he argues, the specific intellectual accrues power through the production of 'truth effects', which sustain and are in turn extended by government interests. From Foucault's perspective, the Lysenko affair can thus be seen not as exceptional but as an example of the wider imbrication of scientific knowledge and power in the twentieth century.

Lysenko was a peasant with very little formal training who started to experiment with plant breeding while working as a technician at the agricultural station of Gandzha (Kirovabad). In 1928 he claimed to have discovered a new technique of 'vernalisation' whereby winter wheat could be made to flower early, so that it could be planted further north than had previously been possible. This was at the time of the forced collectivisation of Russian farms, which was also a period of disastrous harvests and widespread famine. Lysenko's 'discovery' was publicised in *Pravda* and came to the attention of Stalin, who encouraged Lysenko, firstly because he seemed to offer solutions to intractable agricultural problems (some of them actually caused by collectivisation), and secondly because he could be presented as an aspirational figure and emblem of hope for the disaffected peasantry (the fact that Lysenko always wore peasant dress was especially helpful in this respect). Lysenko went on to make further controversial claims, for example, that new plant forms could successfully be created through 'vegetative hybridisation', a form of grafting that did not involve sexual reproduction; by this means he proposed, among other things, that wheat could be transformed into rye. Lysenko's claims were inflated and his grasp of plant physiology was limited, but his extraordinary rise can be explained largely by its historical and political context. In the USSR, science was run centrally, and once Stalin had put his weight behind a particular approach, it was extremely difficult to change course. Moreover, in 1935 Lysenko met I. I. Prezent, a member of the Academy of Agricultural Sciences and an influential figure in the Communist Party. Encouraged by Stalin, Prezent helped Lysenko to develop a genetic theory that offered a direct challenge to Western (neo-Mendelian) genetics. When Lysenko presented this theory at a meeting of the Russian Academy of Agricultural Sciences in 1948, it sparked controversy across Europe, China and America.

'Soviet genetics', as it came to be known, drew on the Lamarckian concept of the inheritance of acquired characteristics.[14] Lysenko claimed that in his vernalisation experiments, the changes produced in winter wheat had been transmitted to the next generation; according to him, this outcome demonstrated that genetic inheritance was not fixed and that the characteristics of plants and animals could be changed by modifications to the environment—in other words, by nurture. Lysenko also argued that Darwin was mistaken in his conception of a 'struggle for existence' within as well as between species: individuals within a species were co-operative, he maintained, and when resources were scarce, individual organisms would automatically sacrifice themselves for the common good. He elaborated on this theory in a conversation with the French biologist Marcel Prenant about his technique of 'close planting', which was implemented on a wide scale in the USSR in 1949. Prenant writes:

> I allowed myself to put a question to him: 'I admit that young trees should be planted in a cluster; they may thus be better protected at first; but is it not necessary to remove some of them after a few years?' 'No,' replied Lysenko, explaining: 'They will sacrifice themselves for one.' 'Do you mean,' I replied, 'that one will turn out to be stronger and the others will weaken or perish?' 'No,' he repeated, 'they will sacrifice themselves for the good of the species'.[15]

These ideas emerged from a long-standing tradition in Russian biological thought, as Daniel Todes has shown in an illuminating discussion that traces the concept of co-operation back to the zoologist Karl Kessler's *On the Law of Mutual Aid*, first published in 1879.[16] Highlighting the connections between biological theories and their broader social and historical contexts, he argues that the Darwinian concept of a 'struggle for existence' made little headway in Russia as it was a country in which 'capitalism was only weakly developed and political supporters of the two most important classes . . . spoke the language of communalism' (36).

Lysenko presented his research methods as aligned with the principles of dialectical materialism and his science as grounded in an understanding of the dynamic interactivity of the material world. In addition, his theories played into the long-standing Marxist preference for theories of 'soft' rather than 'hard' inheritance. These terms were first coined by the British cytologist and geneticist Cyril Darlington, whose work is discussed in the next chapter. Soft inheritance refers to a theory of continuous hereditary change, in which gene and environment are understood to be interacting continuously, and it is associated with Larmarckian views about the inheritance of acquired characteristics. Hard inheritance refers to the theory of stable, fixed genetic inheritance, which is usually associated with neo-Darwinist ideas of genetic determinism. Lysenko's arguments were pitched firmly in the Lamarckian camp, and his non-deterministic view of inheritance and development intersected productively with Marxist ideology, opening up the prospect of, in Nils Roll-Hansen's words, 'mutual reinforcement between improvement of the social environment and improvement of the "gene fund" of the population'.[17] While the concept of stable inheritance could be used to underwrite ideas of 'natural' social hierarchies, non-determinist Lamarckian theories were more readily aligned with a commitment to social progress.

Lysenko's scientific ideas and the Soviet system of power became mutually constitutive, and those who questioned his work were increasingly at risk. When he became president of the Academy of Agricultural Sciences in 1938, a campaign was started to purge Russian research institutes of 'reactionary' neo-Mendelian genetics. The most high profile victim was N. I. Vavilov, a distinguished Soviet geneticist whose laboratory had been visited by Huxley and Haldane in the 1930s and who had, ironically, encouraged the work of the young Lysenko. Vavilov was repeatedly denounced by Lysenko and was sentenced to death in 1942, dying of starvation in a labour camp in 1943.[18] News of these events began to filter out during the war, but it was not until the European publication of Lysenko's report to the Russian Academy in 1948 that the extent of his grip on power and the brutal suppression of Mendelian genetics became widely known. The reaction was immediate, particularly in the UK and France.

Julian Huxley was one of the first to attack Lysenko and his theories in two articles in *Nature* that were subsequently published in an expanded form as *Soviet Genetics and World Science* (1949). Huxley's main target is what he sees as Lysenko's breaching of the 'basic requirements of scientific method' and his repudiation of the 'universal and supranational character of science'. In the first of his articles, he protests in the strongest terms against the subordination of scientific enquiry to politics, arguing that:

> [S]cience is no longer regarded in the USSR as an international activity of free workers whose prime interest it is to discover new truth and new facts, but as an activity subordinated to a particular ideology and designed only to secure practical results in the interests of a particular national and political system.[19]

He takes issue with Lysenko's failure to follow experimental protocols in the vernalisation experiments and states forcefully that neo-Mendelian genetics is 'not just a theory' but an 'autonomous branch of science' incorporating 'a large body of tested scientific fact, and a number of well-validated scientific laws' (941). Huxley's main objective is to defend the concept of the universality of scientific truth, and he dismisses other issues, such as the fact that 'some geneticists in the USSR may have been directly or indirectly "liquidated"' or that 'Mendelism has sometimes been used to justify undesirable theories and actions, such as Nazi Socialist theories', as 'subsidiary' (935). Huxley was supported by Cyril Darlington, the cytogeneticist whose pioneering work on chromosomes had provided much of the evidence for the Modern Synthesis of evolution and genetics.[20] Darlington was a personal friend of Vavilov and was appalled by the treatment meted out to him and other dissenting geneticists. In 1947 he published an outspoken condemnation of the persecution of scientists in the USSR, and in a 1948 BBC broadcast, he spoke of his disappointment about what appeared to be 'indifference to this persecution on the part of our own scientists'.[21] He was joined in this broadcast by J. B. S. Haldane, whose response to the Lysenko affair was equivocal. Haldane was a Marxist who was at this time a Communist Party member, but he was also a committed eugenicist. He was sceptical of Lysenko's claims, but, like many Western intellectuals in the early years of the Cold War, was reluctant to engage in public criticism of the Soviet state.

It was for this reason that, as Diane Paul has shown, he publicly denied that Vavilov had been arrested and had died in custody even though he knew this to be the case.[22] The embryologist C. H. Waddington was the only prominent biologist who spoke out, not in support of Lysenko's science, but of the wider Soviet policy of aligning science with social and political aims. Unlike Huxley and Darlington, Waddington was sceptical of the claim that science could be seen as a disembodied 'universal' practice that transcended national borders and social imperatives. Indeed, he was one of many left-wing scientists who had supported the 'social relations of science' movement that had grown up in Britain in the 1930s. Led by the crystallographer J. D. Bernal, this group had argued that the development of science should be controlled and directed for the public good—just

Figure 3.1 David Low cartoon satirising J. B. S. Haldane's response to the Lysenko affair, 2 December 1948.

as it was, in theory, in the USSR. Accordingly, in an article in the *New Statesman*, Waddington defended the Russian policy that 'all activities [including science] should consciously be part of an integrated movement of human endeavour, in which all men of goodwill know themselves to be co-operating' and went further, arguing that if scientists in Britain wanted to have an influence on matters of policy, they should accept that they too could no longer remain in an ivory tower imagining themselves to be 'gentleman scientists'.[23]

The Lysenko affair put the spotlight on what was becoming a crucial post-war issue: what was the appropriate relationship between science and government? To what extent should the transformative power of science be controlled by the state, particularly if the state was providing the funding for research? In the Cold War context, was it ethical, for example, for scientists to devote their talents to developing biological weapons? Even during the Second World War, many scientists had been uneasy about the subordination of science to national and military ends: this issue is at the heart of C. P. Snow's novel *The New Men* (1954), which draws on Snow's war-time experience to dramatise the conflicting loyalties of a group of British scientists involved in the development of the atom bomb. In the post-war period it was clear that those countries with the most advanced science and technology would wield the greatest power on the world stage, yet the extent to which research should be directed by government remained an extremely contentious issue.

INVASION OF THE TRIFFIDS

John Wyndham's *The Day of the Triffids* was published three years after the Lysenko affair and offers a powerful imaginative response to many of the questions it raised. Wyndham's novels have not yet had the critical attention they deserve, largely because they have been subsumed into the category of the 'cosy catastrophe', the term coined by Brian Aldiss to describe the tenor of much post-war British science fiction. However, Wyndham's speculative fiction is far from cosy, engaging as it does with key issues that preoccupied scientists and the wider public in this period. In *The Day of the Triffids* he dramatises and critiques the contrasting genetic ambitions of Russia and the West during the Cold War. In this respect, the metaphor of the flesh-eating triffids itself works on two levels. First, these plants can be seen as the embodiment of Lysenko's agrobiological theories. The text is explicit about the fact that they are the product of Lysenko's research. We first learn about them from the main protagonist, a plant biologist named Bill Masen who works for a company farming the triffids in the UK. According to him, they were developed in the USSR, where despite the pathological secrecy surrounding biological science it was known that 'under a man called Lysenko' a number of developments had taken place that could have been 'very successful, very silly, or very queer . . . if not all three at once'.[24] The trifffids, we are told, constitute a new species created through vegetative hybridisation: as a Russian scientist explains to Mason, 'I do not say there is no turnip there. I do not say there is no nettle, or even no orchid there. But I do say that if they were all fathers to it they would none of them know their child' (24). In response to environmental changes, these plants have also acquired new and transmissible characteristics, learning to 'walk' by uprooting themselves and to 'talk' through rattling the sticks on their stems. They thus conform to Lysenko's claims about hybridisation and the inheritance of acquired characters and to his arguments about mutual aid. Josella (the only significant female character in the novel) is the first to notice this:

> [T]hey don't bother about their individuals, and the individuals don't bother about themselves. . . . They sort of work together for a purpose the way ants or bees do—yet you could say that not one of them is aware of any purpose or scheme although he's part of it. It's all very queer—probably impossible for us to understand, anyway. They're so *different*. It seems to me to go against all our ideas of inheritable characteristics. Is there something in a bee or a triffid which is a gene of social organization? (209)

On one level, then, the triffids can be interpreted as a fictional representation of Lysenko's co-operative plants. On another, they can be seen as representing the Russian people themselves and as pointing to the biopolitical implications should Lysenko's genetic theories turn out to be correct and to apply to human beings. Lysenko himself never claimed that his work was relevant to human genetics, but Stalin was interested in this possibility, for obvious reasons. If heredity were not fixed, and if desirable characteristics could be created through environmental change and passed on to the next generation, a whole population could, in theory, be 'improved' or uplifted at a stroke. Add to this the principle of 'mutual aid' and the prospect opens up of creating a 'multiple body' of the Soviet population, endowed with superior powers and working collectively to shared ends. The triffids—inscrutable, ever-increasing in number—represent just such a biopolitical threat, and they begin to take over in Britain because of a slip on the 'tightrope' of the balance of power during the Cold War. Satellites are circling the earth loaded with biological as well as atomic weapons, and a substance designed to disable the optic nerve is accidentally released. Through losing their sight, human beings/the British lose their biological advantage, leaving the field wide open to the triffids/the Russians (96).

In his evocation of the British response to this crisis, Wyndham maps the state of play in Western genetics. Shortly after the catastrophe, the few people who can still see converge on the University of London in Gower Street (the home of the Galton Institute) to discuss strategies for survival. A group of civil servants and academics take charge, and the first principle to be invoked is not that of mutual aid, but the survival of the fittest. One of the speakers at the meeting, a professor of sociology named Vorless, argues that the sighted survivors confront a stark choice: either they devote themselves to helping those who have been crippled, or they dedicate themselves to the continuation of the human race. He strongly supports the latter course, invoking the terminology of eugenics to argue that under the new conditions of life, '[w]e can accept and retain only one primary prejudice, and that is that *the race is worth preserving*. To that consideration all else will for a time at least be subordinate. We must look at all we do, with the question in mind: "Is this going to help our race survive—or will it hinder us?"' (99–100, Wyndham's emphasis). A working-class intellectual, Coker, takes the opposite view, arguing that those who can still see have a moral obligation to help the sightless and relieve their suffering; however, all his attempts founder, and eventually he concedes that his efforts have done nothing more than prolong the pain of the weak.

In pursuit of the goal of preserving 'the race', Vorless and his colleagues plan a community in which everything will be subordinated to reproductive efficiency. In biological terms, fewer men than women are needed for reproduction, so it will consist largely of women and will be polygamous. A few blind women will be accepted, as they will have babies who can see, but only sighted men will be allowed to join: blind men cannot work and are now redundant. This is also a community with eugenic ambitions. It is dedicated not just to the survival of the race but also to its improvement: just before Masen joins he is told somewhat threateningly that '[w]e aren't out to reconstruct—we want to build something new and better. Some people don't take to that' (220). It is a utopian community grounded in the principles of the Modern Synthesis of evolution and genetics, and its aims accord closely with the 'evolutionary humanism', which Huxley presented as an alternative to the Lysenkoist world-view in his *Nature* articles. He defined this 'humanism' in terms of a belief that 'man has the duty of carrying the general process of evolution to new heights' (980), and Wyndham's novel ends with a small group of survivors retreating to the Isle of Wight to pursue this goal. By this means they hope one day to overthrow the marauding triffids, although they know that this will be a difficult task, not least because their genetic strategy cannot offer the quick fix associated with Lysenkoism, but involves a breeding policy across generations. Nonetheless, they are optimistic that they will finally succeed in their 'great crusade to drive the triffids back and back with ceaseless destruction' (233). Wyndham here invokes the war-time rhetoric of the plucky island race and translates it into the terms of Cold War biopolitical conflict.

The differences between Anglo-American and Russian science were emphasised and exaggerated for polemical purposes during the Cold War, and in this respect, it could be argued that the Lysenko controversy played a major part in shaping the direction of genetic science in post-war Britain. The scandal over Lysenko certainly strengthened the position of those geneticists (and they were in the majority) who considered genes to be the most important causes of human characteristics. Such genetic determinism was further supported by Watson and Crick's discovery of the structure of DNA, and a gene-centric understanding of development went on to drive the vast majority of the genetic research carried out during the second half of the twentieth century in both Britain and the US. This model of development was top-down and one-way, as is revealed by the metaphors used by its exponents: in this respect Evelyn Fox Keller has analysed the move from the term 'gene action' to the metaphor of the genetic program, with its invocation of computational power and control, and in the later twentieth century to the linguistic metaphor linked with the Human Genome Project, in which genes are figured as codes or digital instructions for making a human being.[25] However, there were several distinguished geneticists and embryologists who continued to argue for the complex interplay between gene and environment at a cellular level and who contended that changes in the environment could create new and heritable characteristics, rather as Lysenko had suggested. The most important of these advocates of 'soft inheritance' was Waddington, who as we have seen in the introduction coined the term 'epigenetics' to refer to the complex unfolding of genetic material in embryonic development. Anne McLaren's work was also significant, as was that of Donald Mitchie, who argued for a two-way relationship between the germplasm (gene) and the soma. However, as Marion Lamb has argued, work in soft inheritance was hindered in the UK for decades for political reasons. She writes:

> [t]he idea that inheritance was soft was associated with Lysenko and the Soviet Union. Although Great Britain didn't have McCarthyism, the Cold War affected British biologists, and it would have been unwise, particularly for young people, to give Lamarckian interpretations to their data.[26]

DYSGENIC PRESSURES

In *The Day of the Triffids*, Wyndham examines the eugenic fantasies stimulated in this period by hubristic claims about the power of genetics to shape human development. However, such optimistic visions were counterbalanced by anxieties about the increasing 'genetic load' carried by modern man. The concept of the genetic load was developed by the American geneticist Hermann Muller, based on the understanding that the human genome was constantly subject to change by mutation. Some natural mutations, Muller argued, were beneficial, but most were harmful and could lead to greater susceptibility to cancer, diabetes, hypertension, other non-communicable diseases and several forms of mental illness. The gradual accumulation of such mutations, spread through reproduction, constituted the genetic load of the population, which was, he claimed, increasing more rapidly as harmful genes were no longer being eliminated in the way that they had been in the past. Medical and social advances, in the form of better hygiene, nutrition, housing and medical care, meant that many of those carrying such genes would survive (and reproduce) as they had not done previously. Moreover, the genetic load was likely to increase further because of the release of atomic radiation: indeed it was Muller's own early work on *Drosophila* (the fruit fly) that had first shown that mutation could be induced by radiation.

Muller's ideas were echoed and endorsed by Cyril Darlington, who argued in *Genetics and Man* (1964) that advances in medicine were ensuring that '[r]aces, classes and individuals of technically less enterprising types, which would not be capable of surviving unassisted, are now multiplying'. In consequence, he claimed, the human race was faced with the danger of 'genetic decay'.[27] As we have seen, Francis Crick also expressed concerns about the dysgenic effects of differential breeding. Julian Huxley claimed in a 1962 Galton Lecture that the progress of medicine had 'permitted numbers of genetically defective human beings to survive and reproduce', leading to the probability of genetic regression and degeneration. This possibility had been 'converted into a certainty' by the mutations arising as a result of atomic and nuclear explosions since the end of the last war.[28] Aldous Huxley made similar points in *Brave New World Revisited* (1959), a book in which by his own admission he is far more pessimistic about humanity's prospects than he had been in *Brave New World* (1932). Echoing his brother's concerns, he asks:

> And what about the congenitally insufficient organisms, whom our medicine and our social services now preserve so that they may propagate their kind? To help the unfortunate is obviously good. But the wholesale transmission to our descendants of the results of unfavourable mutations, and the progressive contamination of the genetic pool from which the members of our species will have to draw, are no less obviously bad.[29]

For Muller the obvious answer to this 'contamination' of the gene pool was positive eugenics in the form of eutelegenesis, which he re-named 'germinal choice'. Eutelegenesis refers to artificial insemination by a 'superior' donor and was an idea first mooted in the 1930s by an English postal clerk, Herbert Brewer. It met with an enthusiastic response from members of the Eugenics Society, including Julian Huxley, who wrote to C. P. Blacker in 1937 about a request he had received to father a couple's child for 'eugenic' reasons. Huxley was keen to do this and to put together a 'eutelegenetic' panel of donors, but Blacker was concerned about donor anonymity, and the idea lapsed.[30] However, Muller argued that improved techniques of artificial insemination, especially the ability to freeze and store sperm over long periods, now offered the opportunity for those with valuable genes to spread them more widely. If 'superior' donors and willing recipients could be found, 'germinal choice' would improve the quality of the gene pool and counter the dangers of the high 'genetic load'. Several eminent geneticists, including Ernst Mayr and Francis Crick, subsequently donated sperm to the Hermann J. Muller Repository for Germinal Choice.[31]

It was Julian Huxley who made the case for 'germinal choice' in Britain, with considerable support from Haldane behind the scenes. For Huxley, such positive eugenics offered the possibility of speeding up 'psycho-social evolution' on a grand scale. In his lecture 'Eugenics in Evolutionary Perspective', he argues enthusiastically that:

> [m]odern science is providing the necessary techniques, in the shape of artificial insemination and the deep freezing of human gametes. The effects of superior germ-plasm can be multiplied ten- or a hundred-fold through the use of what I call E.I.D.—eugenic insemination by deliberately preferred donors—and many thousandfold if the superior sperm is deep-frozen. . . . When deep-frozen ova too can be successfully engrafted into women, the speed and efficiency of the process could of course be intensified.[32]

He goes on to write glowingly of the merits of 'vicarious parenthood' and to suggest that couples who opt for such a process will be richly rewarded by having children who are 'outstanding' in the qualities they themselves admire.

A form of vicarious parenthood is the subject of Wyndham's novel *The Midwich Cuckoos*, which explores some of the ethical dilemmas posed by the emergence of a group of 'outstanding' children. The novel starts with what can be seen as a reproductive experiment writ large, as an invisible, glass-like hemisphere drops down from space onto the village of Midwich, sealing it off from the outside world and rendering all the inhabitants unconscious. While the population is, as it were, anaesthetised, fertilised ova are implanted in all the women of child-bearing age. The women thus become surrogate mothers of children who turn out to be extraordinarily gifted—a recurring trope in science fiction at this time. Physically, they develop far more rapidly than ordinary children and are as advanced at the age of nine as normal sixteen year olds. They are equally advanced mentally, especially as they work collectively through a kind of thought transference (echoing the triffids in this respect), so that their intelligence is shared and multiplied. We see the children principally from the perspective of Gordon Zellaby, an upper-middle-class novelist who has read widely in evolutionary and genetic theory. He understands them as representing a new stage of human evolution and offers a range of possible explanations as to how this might have come about. Initially he thinks that they might represent a natural evolutionary shift, prompted by the fact that human beings have come up against the limits of their developmental potential. They might, accordingly, be mimicking the strategy adopted by insects in order to get round such a problem: '[t]he laws of physics prevent them increasing in size, so they contrive greater efficiency by acting as a group. We ourselves combine in groups consciously, instead of by instinct, for the same purposes . . . why shouldn't nature produce a more efficient version of the method?'[33] Alternatively, in a challenge to Darwinian evolutionary theory of gradual change, he suggests that the children might constitute a new race of humankind that has 'popped into existence fully formed, or at least with no clear line of ancestral development' (204). Here he seems to anticipate Stephen Jay Gould's theory of 'punctuated evolution', itself based on an article published by Ernst Mayr in 1954, which Wyndham might well have known.[34] Finally, on the principle that nothing should be ruled out, he entertains the idea that the children may have been created by extra-terrestrial beings who are using the world as 'a rather disorderly testing-ground' where they let loose new kinds of being and observe the struggle for survival between old and new life forms (205).

The novel can be read as a thought experiment in which the placid, superlatively ordinary community of Midwich is confronted by what is effectively a super race, a more advanced version of humanity that represents a clear biopolitical threat to the villagers, as Zellaby indicates when he describes the children as an enemy within. The children share Zellaby's understanding of the situation. As one of them explains, the case is very simple for '[i]f we exist, we shall dominate you—that is clear and inevitable. Will you agree to be superseded, and start on the way to extinction without a struggle? I do not think you are decadent enough for that' (198–9). Zellaby decides that it is 'our duty to our race and culture to liquidate the Children, for it is clear that if we do not we shall, at best, be completely dominated by them, and their culture'. Accordingly, he blows them all up, sacrificing his own life in the process (208). The novel thus offers an unusual perspective on biological racism, which in this case legitimates the 'liquidation' not of an inferior but of a superior race. The children have similarities to the Nazi Aryan elite, as they are tall and luminous, with 'lucent skin' and golden hair; as 'potential geniuses' who have been educated separately, they also offer hints and shadowy reflections of the meritocracy. They mirror the eugenic ideology of Nazi Germany and the eugenic ambitions that continued to circulate in Britain, but through the violent climax of the novel, Wyndham draws out the ambiguities surrounding the fascination with a super race that marked science fiction of this period. The 'higher' types of human beings imagined by writers ranging from Arthur C. Clarke to Doris Lessing come out of a profound disillusion with humanity and what it is capable of, as demonstrated by the Holocaust and the atomic bomb. In delineating new species that yet retain recognisably human characteristics, these writers seem to be exploring the possibility of a redemptive future that will be brought about by human intervention in the processes of evolutionary change. Yet as Wyndham's novel suggests, the development of a 'higher' type of human will necessitate the elimination of the old, a price that Zellaby, at least, is not yet willing to pay.

GENETIC MUTATIONS

Wyndham's *The Chrysalids* explores a further dysgenic pressure, the threat posed by atomic radiation. The fact that radiation could induce genetic mutations had been known as early as the 1920s, and after the bombing of Hiroshima and Nagasaki, both the short and long term health effects of atomic radiation were key issues for both the British and American governments. In 1946, the American National Research Council established the Atomic Bomb Casualty Commission to study the biological effects of the bombings in Japan, and between 1954 and 1964, it set up a number of other committees to review the effects of radiation. In the UK, the Medical Research Council was designated as the lead body to advise the government on this question, producing a report, *Hazards to Man*, in 1956. The World Health Organization also convened a study group of distinguished geneticists, including Muller and Penrose, to discuss the effects of radiation on human heredity (Figure 3.2). In its 1957 report, this group concluded that the damage produced by radiation was real, and that there was a need for extensive research into the harm caused by both peaceful and military uses of nuclear energy 'with a view to safe-guarding the well-being of future generations'.[35]

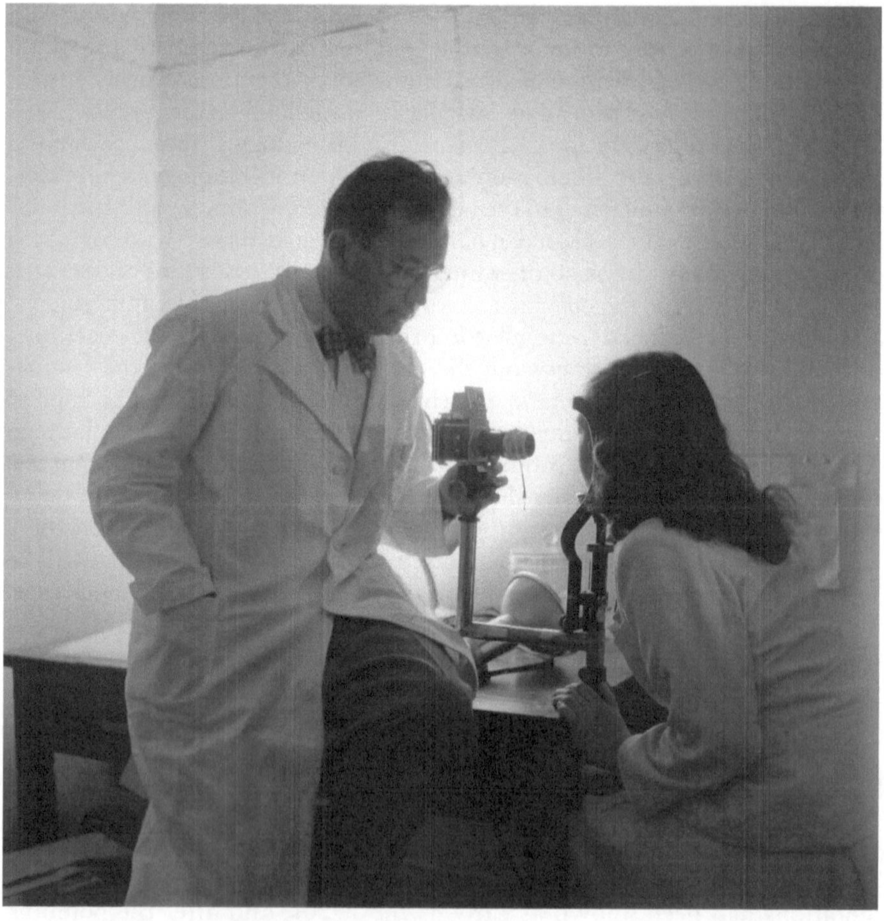

Figure 3.2 A Hiroshima survivor being examined at the Atomic Bomb Casualty Commission Clinic, 1 October 1949.

The main concern was over long term genetic damage, which was often represented as an insidious threat that was all the more dangerous because it was invisible. In Peter Alexander's 1957 Pelican paperback *Atomic Radiation and Life*, for example, he warns that:

> The apparent normality of children born to survivors of the Hiroshima and Nagasaki atom bomb attacks . . . does not invalidate the dose levels for genetic damage which have just been deduced. One of the reasons is that very few mutations are dominant enough to be immediately recognised when inherited from only one parent. *They do, however, add to the genetic load, and payment will be exacted at some stage in a future generation.*[36]

As this near-Biblical language suggests, the concept of genetic mutation carried a powerful cultural charge in this period, reflecting deeply felt fears for the future. The abundance of science fiction texts and films that focus on the trope of the mutant child is further evidence of this anxiety, which has been ascribed to guilt about the war by David Seed, who argues that in mutant narratives, 'a generation displaces responsibility for the Bomb on to the babies'.[37] Foucault's discussion of atomic power in *Society Must Be Defended* opens up a further perspective on such anxieties. Foucault understands atomic power as an expression of biopower pushed to its paradoxical limits. He argues that the power exercised through the manufacture and use of the atom bomb is 'exercised in such a way that it is capable of suppressing life itself. And, therefore, to suppress itself insofar as it is the power that guarantees life'.[38] Cyril Darlington makes this point more bluntly in *Genetics and Man* when he writes that '[e]quipped with new types of explosives which depend upon splitting the atom, man is able for the first time to make war by breaking the chromosomes and exploding heredity—for that, as we have seen, is the way atomic radiation kills' (350). He refers here to the fact that atomic radiation targets the most rapidly dividing cells in the body, producing radiation sickness by attacking bone marrow, gut and skin cells and creating genetic mutations by damaging DNA. To the extent that radiation intervenes directly in the process of reproduction by damaging germ cells, destroying or modifying *future* life, it takes biopolitical warfare to a new and devastating level.

Wyndham's *The Chrysalids* (1955) explores the aftermath of precisely such biopolitical warfare, conjuring up a post-atomic landscape in which large areas of the world have been decimated while in others plants, animals and humans have mutated in fantastic and monstrous ways. In response to this man-made catastrophe, a religion is established that deflects the blame for it onto the mutants, while reinstating the Christian belief in man as created in the divine image. Religion and biopower are fused in a regime of biological fundamentalism, in which all mutations, but especially those in human beings, are abhorred and eliminated. The home of the young narrator, David, contains exhortations along these lines:

> The one on the left of the fireplace read: ONLY THE IMAGE OF GOD IS MAN. The one on the right: KEEP PURE THE STOCK OF THE LORD. On the opposite wall two more said: BLESSED IS THE NORM, and IN PURITY OUR SALVATION. The largest was the one on the back wall, hung to face the door which led to the yard. It reminded everyone who came in: WATCH THOU FOR THE MUTANT![39]

As Foucault suggests, the concept of the norm is crucial to the workings of biopower, mediating between the body and the population, between disciplinary and regulatory power.[40] However, in the fictional world of *The Chrysalids*, biopower goes beyond what is acceptable in liberal democracies and is closer to what Foucault terms the 'paroxysmal' logic of National Socialism, in which biopower and the right to kill are joined together and transmitted throughout the entire social body. For Foucault, biopower in the Nazi state is murderously generalised:

> Ultimately, everyone in the Nazi State had the power of life and death over his or her neighbors, if only because of the practice of informing, which effectively meant doing away with the people next door, or having them done away with.[41]

Similarly in Wyndham's novel, the kingdom of Waknuk is maintained in a state of biological purity by vigilant inspections and by informing, often under torture. David himself betrays his friend Sophie (who has six toes) after he has been brutally beaten. He suffers terrible guilt over this, but his mother feels none when she causes the death of her sister's abnormal baby, for she has internalised the view that such creatures are 'abominations' who must be sterilised before being left in the wild to perish. Wyndham's focus on the elimination of vulnerable children ensures that the reader's sympathies lie with the mutants, and the progressive, relativist strand of the novel is further reinforced through the character of David's uncle Axel, who offers a trenchant critique of the notion of 'the norm'. Axel has travelled widely and explains to David that there are many different kinds of humans in the world, with all sorts of mutations; he points out that it is absurd that each should imagine that it is only their own type which is the 'true pattern' or image of God.

Yet while offering a critique of the concept of the norm, the novel shares with *The Midwich Cuckoos* a fascination with the trope of a super race, which Wyndham links directly with atomic radiation. The suggestion is that radiation has created a 'superior variant' of people who (like the Midwich cuckoos) can 'think-together' and 'apply the composite team-mind' to problems. David is one of them; he realises when he is quite young that he is telepathically gifted, and with Axel's help he also sees that this will be viewed as an 'abomination' within his community. Eventually he and a small group of children who have this gift, including his sister Petra, are discovered and leave Waknuk pursued by vigilantes. Petra, however, turns out to have a mind so powerful that she 'can throw a thought halfway round the world' and she is able to make contact with some 'Sealanders' (New Zealanders) who share her extraordinary talents. They fly to the rescue, descending like avenging angels; specifically, Aryan angels, for their most notable characteristic is whiteness. Their leader is described in these terms: 'More than anything it was the lightness of her face that made us stare. It was not pallor, it was simply fairness, like new cream, and with cheeks that might have been dusted with pink petals'. They are also 'new and perfect' and seemingly ageless—their leader's face has 'scarcely a line in its smoothness. . . . We found it hard to believe that any real, living person could look like that. . . . For she was no girl' (192). In short, the Sealanders are uncanny, reflecting Wyndham's ambivalent attitude to human 'improvement', which is also evident in the final scene of the novel.

After their rescue, the children realise that the cost of their escape has been the destruction of every living thing around them: 'In the pause one became aware of the uncanniness of the silence all about us. There was not a sound to be heard now. Not a movement. Even the leaves on the trees were unable to rustle' (194–5). The Sealanders justify this destruction in Darwinian terms, arguing that 'the essential quality of living is change; change is evolution', and that just as the dinosaurs passed away, so 'normal' human beings must give way to the new 'superior' variant they represent. David acknowledges the logic of the argument but is also repelled by it, remembering the pain of his aunt's suicide and Sophie's death. Nonetheless, the novel ends with his flight to Sealand and his enchanted recognition of it as the land that haunted his childhood dreams and was explained to him then as an echo of the world as it had once been. In drawing together visions of a time before the fall and of a redemptive future, the text creates a narrative of progress informed by the kind of evolutionary humanism that Huxley had called for in his critique of Soviet science. As we have seen, Huxley argued that 'man has the duty of carrying the general process of evolution to new heights . . . in discharging that duty rightly he will be providing and ever expanding new possibilities of fuller living for future generations'. In this novel, Wyndham fleshes out this ideal, while at the same time drawing attention to the exclusionary violence latent within it (980).

An upsurge of post-war, post-atomic fiction sought to re-imagine 'the human' in this way, ranging from pulp science fiction films to the 'literary' science fiction of Doris Lessing and Naomi Mitchison. Many of these texts engage with the notion of a 'higher' type of human being, and such 'new' humans frequently turn out to be both extremely intelligent and telepathic. The emphasis on intelligence may be related to the fetishisation of IQ in post-war Britain, while the interest in telepathy could be linked to the fact that both the CIA and the Soviet intelligence services were rumoured to be carrying out experiments in telepathy in the 1950s and 1960s. In a 1969 interview with Studs Terkel, Doris Lessing gives credence to this, insisting that 'different forms of extrasensory perception are being seriously researched and accepted. And have you ever really thought about how the atmosphere's changed about something like telepathy in ten years? Of all places it was the Soviet Union that suddenly made the announcement that they were experimenting into the use of telepathy for space travel'.[42] This was also the year in which Lessing published *The Four-Gated City*, the novel that paved the way for her 'space fiction' and her engagement with genetics and eugenics.

INTER-GALACTIC EUGENICS

In the second volume of her autobiography *Walking in the Shade*, Doris Lessing offers an account of the following incident when she was staying with the novelist Naomi Mitchison:

> Naomi asked me to take a certain inarticulate young scientist for a walk. 'And for goodness' sake, get him to say *something*—his tongue will atrophy.' His name was James Watson. For about three hours we walked about over the hills and through the heather, while I chatted away, my mother's daughter: one should know how to put people at their ease. At the end of it, exhausted, wanting only to escape, I at last heard human speech. 'The trouble is, you see, that there is only one other person in the world I can talk to.' I reported this to Naomi, and we agreed that it was as dandified a remark as we could remember, even from a very young man. Quite soon he and Francis Crick would lay bare the structure of DNA.[43]

This anecdote points to Lessing's awareness of, and interest in, contemporary biological science. While Lessing's explorations of gender and colonialism have received a great deal of critical attention, her interest in twentieth- and twenty-first-century science has been neglected, even though it underpins much of her most significant work. Her friendship with Mitchison was crucial to this aspect of her writing, as it was Mitchison who introduced her to science and scientists through the 'salon' established at her country house at Carradale, to which an eclectic range of people were invited and where, in Lessing's words, 'the famous divide in the culture between science and the arts did not exist'.[44] Mitchison's guests were especially likely to include biologists: her brother was J. B .S. Haldane, and her son, Avrion Mitchison, was the up-and-coming young zoologist who had invited James Watson to stay. Lessing describes the scientific talk she heard there as 'irresistible'.

It is likely that it was Mitchison who inspired Lessing to turn to science fiction in the 1970s, a development that confounded many of Lessing's readers and critics. However, her move in this direction becomes more intelligible if her space fiction is understood as part of a British tradition of scientifically informed, speculative fiction, which begins with H. G. Wells and goes on to include J. B. S. Haldane's *Daedalus* (1924), Olaf Stapledon's *Last and First Men* (1930), Aldous Huxley's *Brave New World* (1932) and Mitchison's *Memoirs of a Spacewoman* (1962). The parallels between Lessing and Mitchison's space fiction are striking. Both engage with the models of human development opened up by genetic science, and both explore their implications for thinking about both the pre-history and the longer term future of the species. Lessing is particularly concerned with the question of genetic mutation in an evolutionary context, and in *The Four-Gated City*, she, like Wyndham, explores the possibility of genetic mutation caused by atomic radiation, which she presents in almost entirely positive terms. From early in the novel, there are hints that a large-scale radioactive incident is imminent, but when it happens, its destructiveness is cancelled out by the fact that it has kick-started a new phase of the evolutionary process, with humanity shifting up a gear and moving to a higher level. Lessing emphasises too that there are already in existence many 'forerunners' of this new phase, including the main protagonist of the novel, Martha Quest.[45]

In *The Four-Gated City*, Martha's journey to enlightenment is constructed with reference to the discourse of anti-psychiatry as well as that of evolutionary biology. In accordance with R. D. Laing's theories that 'madness' can represent a form of sanity and that a journey into madness can be a transformative experience, Martha deliberately induces a mental breakdown.[46] After her recovery, she walks out into the street and looks up at the crystalline beauty of the sky, then down to the earth where she is confronted by 'an extraordinarily hideous creature, who stood watching her, out of eyes that were like coloured lumps of gelatine that had fringes of hair about them and bands of hair above them, and which half protruded from a bumpy shape of pinkish putty, or dough-like substance'. [47] Martha's experience of transcendence leads her to see her fellow human beings as physically repellent, lumpish creatures. Through language that emphasises decay and degeneration, Lessing suggests Martha's powerful feelings of revulsion from the people around her and her efforts to abject and repudiate them. For Martha human beings are physically and mentally 'defective', a term that is repeated insistently; the people around her are 'defectively evolved animals' whose 'eyes were half-useless: many wore bits of corrective glass over these spoiled or ill-grown organs; their ears were defective . . . their mouths were full of metal and foreign substances to assist teeth that were rotting' (555). Martha's 'enlightenment', then, produces a sense of separation that is constructed in explicitly eugenic terms. It is as though, to borrow Foucault's terminology, a break is being inserted into the species that separates those who are more highly evolved, like Martha, from the ordinary people around her with their 'atrophied' faculties (554–6).

In this respect, Martha comes to see herself and her friend Lynda Coldridge, who has telepathic powers, as 'experimental models' for the development of the race. They have capacities above the normal and are 'in the main line of evolution', that is, they are representatives of the evolutionary future. They are also closely aligned with the 'new children' who are born after the atomic disaster. These children are extremely diverse, with a range of talents and attributes. Some appear to be 'idiots', but they can communicate with each other without speech. Another group can tune in to other dimensions and have telepathic powers far stronger than Lynda's; others can 'see' pictures from what seem to be other worlds. For Martha, the most important group are the 'new children' who are defined in terms of a new stage of evolution that incorporates and surpasses all that has gone before: '[t]hey don't have to be shielded from the knowledge of what the human race is in this century—they know it. . . . It is as if—can I put it like this?—they are beings who include that history in themselves and who have transcended it' (703). The children have 'a gentle strong authority' and act as 'guardians' both for the other children and the adults. Martha concludes that 'Nature has had enough of us' and that 'one day all the human race will be like them' (703). The social and political contradictions explored in this long and complex novel are thus resolved by means of a eugenic fantasy in which a 'superior' species with 'a comprehension we can't begin to imagine' takes control and, by implication, redeems humanity.

Lessing's space fiction quintet *Canopus in Argos* enlarges on these themes. The first novel *Shikasta* (1979) is, like *The Four-Gated City,* rooted in the framework of the Modern Synthesis of evolution and genetics. The balance of life on Shikasta (earth) is presented in strictly Darwinian terms: 'every life-form preys on another, is supported by another, and in turn is preyed upon', and the post-Darwinian concept of 'degeneration' is frequently invoked in an account of the planet's history.[48] The language of genetics is also crucial, as the uneven development of humanity is interpreted in terms of the mixing of genetic stocks, the emergence of 'good genetic material' and the preservation (or loss) of 'genetic memory'. However, Lessing also moves to locate the Modern Synthesis within a wider context of 'cosmic evolution', a term that for her has at least two senses. First, it invokes the insights of twentieth-century physics and its understanding of the earth as part of a cosmos that is billions of years old and billions of light years in extent; Darwinian evolutionary time is thus subsumed into a vaster history and time frame. However, Lessing also uses the term in the sense in which it was used by Idries Shah, a populariser of Sufi philosophy. Lessing was a pupil of Shah and was attracted by his theory that cosmic evolution depends on the conscious development of the higher potential of humanity.[49]

In the *Canopus in Argus* novels, Lessing explores the interplay between the inter-galactic empires of Canopus, Sirius and Puttiora. Science fiction, with its central trope of the 'stranger in a strange land', has frequently been used to explore the dynamics of colonialism and post-colonialism, and in this respect, it was the obvious genre for Lessing to choose for her return to the colonial themes of *The Grass Is Singing* and the earlier novels in the Martha Quest series.[50] However, a number of critics have suggested that in the representation of empire in her space fiction, Lessing re-instates the colonial assumptions that were so trenchantly critiqued in her earlier work. More specifically, I would argue that in her space fiction she isolates and re-inscribes in *positive* terms the eugenic assumptions that are implicit in the racism viewed so critically in her earlier work. In the *Canopus* fictions, she represents an imperial hierarchy that has eugenics at its heart, and the narrative appears to endorse the philosophy of an empire grounded in eugenic ambitions. The dominant empire, Canopus, is presented as a benign presence and a force for good even though its whole logic and rationale is eugenic. The Canopeans are described as 'higher, more finely tuned organisms, in touch with forces far beyond, far higher than most creatures could ever dream of', and they see their mission as developing and enhancing the other species inhabiting their territories. The techniques they use are genetic, involving careful selection and hybridisation in order to develop species possessing the attributes they prize, which are intelligence, empathy and an ability to 'think together'. Despite Lessing's inventive depiction of these new and supposedly 'alien' species, they remain recognisably human; as such, they can be read as a series of variants on an imaginary super race.

Shikasta traces the history of the earth over millennia, starting from the point when Shikasta (the earth), a planet the Canopeans are monitoring, is affected by a burst of cosmic radiation that causes the rapid development of a 'promising' species:

> They were already three and four times the height of the little creatures who were their ancestors. They were showing tendencies to walk upright. They showed rapid increases in intelligence. Our experts told us that these creatures would continue a fast evolution and could be expected to become a Grade A species. (28)

Canopus decides to implement an 'all-out booster, Top-Level Priority, Forced-Growth Plan' to speed up the evolution of these proto-humans, the mechanism for this being the introduction of a species of giants from another planet to tutor the 'ex-monkeys' (8). The relationship between the two species is symbiotic, and as they are genetically incompatible, there is in this case no inter-breeding. However, the plan is successful, with the natives learning new skills and gaining in intelligence and life-span, and a 'Lock' is therefore established between Canopus and Shikasta. Here another of Lessing's preoccupations comes into play, as the Lock depends on the power of telepathy, which enables 'the free flow of thought, ideas, information, *growth* between planet and planet' (33). As we have seen, Lessing is a firm believer in telepathy and an advocate of the group mind, explaining its significance in the preface to *The Sirian Experiments* when she writes that:

> We see ourselves as autonomous creatures, our minds our own, our beliefs freely chosen, our ideas individual and unique . . . with billions and billions and billions of us on this planet, we are still prepared to believe that each of us is unique, or that if all the others are mere dots in a swarm, then at least *I* am this self-determined thing, my mind my own. . . . How do we get this notion of ourselves? It seems to me that ideas must flow through humanity like tides.[51]

As discussed in Chapter 5, telepathy can be a powerful metaphor for a progressive re-imagining of subjectivity and inter-subjectivity. However, Lessing's comments here seem closer to a eugenic perspective in which individual rights and freedoms are subordinated to the need to develop or preserve the race. In this respect, her description of excessive individualism as 'the degenerative disease' is telling.

Canopus's eugenic policies include 'boosting' other populations with Canopean genes, a policy that is implemented on Shikasta when a second cosmic accident damages the Lock. The Canopeans represent this intervention in terms of a religious rite, as selected 'superior' Shikastans are given the privilege of mating with 'sacred beings' who have come from a higher sphere. For the Canopeans—as for twentieth-century eugenicists—breeding is indeed a sacred duty, only to be undertaken with great care and forethought. Reproduction is their religion, as is suggested through the representation of eugenic marriage, which assumes great significance in *Shikasta*. As Johor, a Canopean representative on Shikasta, explains to his lover Sharma, marriage should always be approached in a holistic way, with a view to developing *all* the highest qualities of the species: intelligence, health and insight. Marriage is also a matter that goes beyond individual interests and desires as it is crucial to the vitality and development of the race. Looking back to the planning of marriages in the past, at the high point of Shikasta's development, Johor describes the way in which individuals and families approached each other. In Johor's ideal scenario, individuals see themselves primarily as representatives of their race:

> [w]hat they are thinking of is how they can carry the best they know into the future of the race, their culture. They see themselves, feel themselves, as the bearers of culture. Yes, they discuss family characteristics—this is a good family, the mother is good and balanced and beautiful enough, and the father is also these things, and his line too. When these young people know these things are being discussed, it is not with a sense of personal affront, which is how we would, now, in these days, experience a discussion about—not our wonderful and precious selves—but our importance as representatives. . . and all the time their children are implicit in what they say and think and do, for the knowledge of how to keep a strong, continuous healthy civilisation is the deepest thing in them. (353)

He goes on to align such eugenic marriages with a broader cosmic harmony, telling Sharma that in the past, partners chose each other because 'they were born with the knowledge of *how* to do this—because they were complementary, and this was judged always by how they stood in relations to stars, planets, the dance of the heavens, the forces of the earth, the moon, our sun' (356). This is not just rhetoric, as Johor goes on to marry and have children not with Sharma, whom he loves, but with the sensible and determined Suzannah. This eugenic decision is vindicated when at the end of the novel, after Shikasta's third world war, it is these resourceful children who survive and are ready to re-build the planet.

In the third of the *Canopus* novels, *The Sirian Experiments*, Lessing explores the consequences of a misapplication of eugenic principles in an empire that lacks what she presents as the 'fuller' understanding of Canopus. In this respect, the novel can be read as a critique of a purely instrumental approach to eugenics. The Sirian empire is advanced technologically, but its eugenic experiments are strictly utilitarian, to the point in some cases of downgrading rather than advancing other species. The story of the Loombis is a case in point. Sirius breeds them to act as a low-grade workforce, a reservoir to be used for 'ordinary, heavy, undifferentiated work'. Their story is told at some length to illustrate the damaging effects of Sirian exploitation, which results in biological changes, as a race that was once 'pleasant, easygoing, unsuspicious' becomes 'nervous and paranoid' (46). The Sirians, then, unlike the Canopeans, subjugate and enslave less 'advanced' species; they also carry out 'biosociological' experiments on them without regard for their mental or physical health. Under the tutelage of a Canopean teacher, a representative of Sirius comes to a painful realisation of the damage caused by Sirian colonisation and experimentation. She also begins to understand the Canopean perspective on the wider balance of forces in the universe. In conversation with her Canopean guide Klorathy, she learns that Canopus has implemented eugenic policies and has 'destroyed cultures that have become corrupt' (267). However, he maintains that:

> when we have been forced to use such means, in order to keep our balances within measure, then these have been small local cultures. A city . . . a group of two or three cities . . . even a few particularly damaging individuals. At this very time, in the area of the great inland seas. . . . And he seemed distressed, in pain . . . we are being forced to take certain steps. (268)

Klorathy tries to make a firm distinction between the utilitarian eugenics of Sirius and enlightened Canopean eugenics. The latter, he contends, are always and only implemented in accordance with the laws that are 'inherent in the nature of the Galaxy, of the Universe'. A further principle he invokes is that of the 'necessity', which is linked to the question 'what is a genus *for*? What is its function? What does it do? What part does it play in the cosmic harmony?'

(273). Klorathy and the Canopean representatives in *Shikasta* are presented as the most admirable figures in Lessing's imaginary universe, and they are the ones with whom the reader is invited to identify most closely. Accordingly, it is difficult not to read the novels as endorsing their beliefs, which fuse together Shah's theories of cosmic evolution and Huxley's evolutionary humanism. Their philosophy also resonates interestingly with that of Francis Galton, who positioned 'men and all other living animals' in the context of an 'extended system of cosmic action' when making his case for eugenics (45).[52] For both Galton and Lessing, the idea of cosmic harmony is invoked to cover over and legitimate a commitment to coercive eugenic strategies.

MEMOIRS OF A SPACEWOMAN

Naomi Mitchison had close connections with almost all the key players in twentieth-century genetics. The daughter of the well-known physiologist J. S. Haldane and the sister of J. B. S. Haldane, she describes in her autobiography having grown up always with 'my father's lab in the background'; she also describes the genetic experiments she and her brother carried out on guinea pigs when they were children.[53] As Lessing attests, she knew James Watson well, and she was also good friends with the embryologists C. H. Waddington and Joseph Needham. *Solution Three*, the second of her science fiction novels (discussed in Chapter 5), is dedicated to James Watson, while *Memoirs of a Spacewoman* (1962) is dedicated to Anne McLaren, an embryologist who was at that time working at Waddington's Institute for Animal Genetics.[54] In an essay she contributed to a book edited by Mitchison, *What the Human Race Is Up To*, McLaren outlines the epigenetic view of development, which she shared with Waddingon, and she also highlights her interest in maternal influences on development. She writes that:

> [t]he resemblance between parent and offspring may be brought about in various ways: through, for example, the chromosomes of the egg or sperm nucleus; or through some other part of the egg or sperm cell; or in the case of mammals, through the intimate relationship between mother and young during gestation.[55]

She goes on to make the neo-Lamarckian point that when more is known about non-genetic inheritance, 'we may find that it is sometimes possible to influence inherited characters by manipulating the conditions of life of the animals or plants' (171).

In *Memoirs of a Spacewoman*, Mitchison engages with these ideas and those of Waddington, although not quite in the enlightened way that Susan Squier suggests when she argues that the text prompts us 'to rethink our definitions of self and other' in the context of a feminist critique of science (186).[56] The novel is set thousands of years into the future, and the main protagonist, Mary, is a scientist and an explorer who has been trained in communicating with alien species. On two occasions when she is not travelling, she volunteers to host an alien graft in a kind of cross-species pregnancy. In the first experiment, Mary develops strongly protective feelings towards her graft, worrying about its safety and 'feeling it was part of me', to the point where she experiences a form of post-natal depression when the graft separates from her.[57] The graft influences her strongly, stimulating a longing for water, for example, but what the experiment also demonstrates, in line with McLaren's work, is the 'intimate link between mother and child during gestation'—even though the graft is biologically unrelated to Mary. After it separates off, it is found that it has acquired her taste in food and also in music, being soothed by the music she played during pregnancy. Here Mitchison, inspired by McLaren, anticipates current research into the way in which 'maternal effects' prepare the child for its post-natal environment.[58] The second experiment is more disturbing, however, literalising the trope of the alien within, which often acts as a metaphor for loss of self in pregnancy. This time Mary becomes so attached to the graft that she is possessed by it, to the point where she is unable to recognise herself *as* herself in a photo taken at the time. She ceases to be 'a civilized scientist' and demands sex from her lover while being tormented (again) by a desire for water and bathing. In all of this, she learns later, she is following the biological imperatives of the graft, as in order to reproduce, it needs to 'deliquesce' and infiltrate her fertilised egg. Mary escapes this fate, unlike another mammal in the same experiment, which gives birth to 'eight dreadful little objects, each of which was like a slightly developed foetus swollen out with a graft-like texture' (155). As this description suggests, the text does not endorse such cross-species breeding, which results in defective and monstrous beings; rather, the old eugenic ideal of 'breeding true' is vindicated.

The theme of species purity is further developed in an episode in which Mary and her colleagues encounter a kind of giant butterfly on a far-distant planet. The butterflies fall into two categories. Most of them reproduce through what Mary describes as 'a prolonged sexual wallow' in bogs of algae. The caterpillars are not male or female in any recognisable sense, and the process of fertilisation is described in terms of random and indiscriminate sexual contacts. After the 'bog wallowings', the caterpillars devote themselves to making elaborate patterns out of their own excrement, which comes out of their bodies in 'pellets of several dark and shining colours' (89). Yet when they indulge in these activities, they are attacked by adult butterflies who project 'anger and judgement' onto them, leaving them in a state of abject misery. The butterflies also intervene when the caterpillars hatch out from their chrysalises, ruthlessly destroying any with malformations. Mary observes them as they swoop down onto one victim:

> They were killing it, withering and blasting it, by no physical violence, but by the burning of their blame, which I could feel pouring by me.... It

> lost confidence, lost all will to be born and live. It died. Then I saw that its wing mass was uneven. If it had lived, it would have been with a crippled wing. But it had not lived. Not at least in its new life. (98)

When Mary is able to establish communication, it becomes clear that the butterflies are acting as eugenicists for their species. They believe that the caterpillars' pattern-making constitutes a fault in their developmental processes, as energy that should be directed to 'the area of future wing development' is dissipated in external play, leading to malformations: hence their attempts to discourage it. More strikingly, the reason for their attacks on the 'wallowings' is that they are trying to direct the caterpillars away from this kind of reproduction. They believe that a form of asexual reproduction has now become available to their species and that if the caterpillars can resist the 'wallowings', they will hatch out into a more advanced form of butterfly, which they believe to be immortal. Mary and her team are divided in their reaction to these claims. Mary endorses the eugenic logic of the butterflies and argues against any attempts to interfere with their behaviour, on the grounds that if the butterflies are right about the pattern-making and the wallowing, stopping their interventions could mean that 'a whole generation of malformed butterflies might emerge, crippled in the chrysalis, unable to be part of a butterfly life' (100). Mary also supports their attempts to move to the path of asexual reproduction, believing that they are correct in thinking it will lead to the development of a 'superior' type. This new species is described in terms strongly reminiscent of the 'new children' in Lessing's *The Four-Gated City* and the Canopeans in *Canopus in Argos*. Both writers use the trope to depict an ideal defined in opposition to the existing human condition, a point that is made explicit in Mary's description of her encounter with the 'immortal' butterflies:

> It resembled the dream which I believe most Terrans have in which it has at last become apparent that all problems are ludicrously simple and solvable, once some principle is grasped: that complete and eternal happiness and knowledge are within easy reach. It is, of course, a negation of the human condition, though something to which perhaps we have struggled a little way. (115)

For Mary, these butterflies represent an inspiration and an ideal, but another member of the team, Françoise, sees them as ruthless exterminators and tries to defend the caterpillars from their attacks: eventually she becomes so enraged by the butterflies' actions that she kills one of them. When the expedition returns to Terra, her punishment is to lose the privileges of belonging to the explorer caste, whose lives in many respects reflect those of the immortal butterflies, as in this imagined future human beings have also begun to separate into two distinct species. On the one hand is an elite of gifted and highly educated explorers who live for millennia because they experience a time 'black-out' during space travel. This elite reproduces rarely and they breed among themselves on eugenic lines. The mass of the population, on the other hand, lead much shorter lives devoted to routine technical work. In consequence of her crime, Françoise will have to live among 'ordinary' people and will age quickly, her fate replicating that of the caterpillars she had sought to defend.

Mitchison was a life fellow of the Eugenics Society, and in *Memoirs of a Spacewoman*, as in her other science fiction, she extrapolates from trends in genetic science to depict a range of possible scenarios in relation to the biological future. Her writing is infused by her extensive knowledge of evolutionary genetics, and she engages in great detail with contemporary debates over the relative influences of genes and environment. Ultimately, however, her vision is rooted in a belief in the possibility and desirability of manipulating reproduction to speed up the development of the human race. At first sight it is perhaps surprising that Doris Lessing, as a colonial outsider, should share this vision, but the twentieth century was, as Fox Keller has so convincingly argued, the century of the gene, a perception augmented by Hilary Rose's comment that it was, equally, the century of eugenics.[59] What Rose calls 'eugenetic' assumptions were deeply embedded in Western culture: the next chapter explores the intersection of these assumptions with the newly prominent issue of race in post-war Britain.

4 Race and the Body Politic

> In view of the admitted existence of some physically expressed hereditary differences of a conspicuous nature, between . . . races, it would be strange if there were not also some hereditary differences affecting the mental characteristics which develop.[1]

> In general the differing genetic endowments of people from different races . . . begin to become apparent, in the anatomical sense, in foetal life. Subsequent development finally produces persons who differ from one another to an extent which is obviously very substantial.[2]

> I think it likely that more than half the difference between the average I.Q. of American whites and Negroes is due to genetic reasons, and will not be eliminated by any foreseeable change in the environment.[3]

The concept of 'race' that developed in the eighteenth century was based on the premise that humanity could be divided into a limited number of distinct, physically differentiated groups, an idea that was taken forward in the nineteenth century through the discourses of anthropology and evolution.[4] The science of anthropology was intimately bound up with the processes of European colonisation and reached a peak in the early-nineteenth century, as scientists were encouraged to record physical differences between themselves and the 'new' people they encountered. However, as Foucault points out in *Society Must Be Defended*, evolutionary theory was also crucial in the construction of the modern concept of race.[5] In *The Descent of Man* (1871), Charles Darwin argued that there were significant biological differences between races, though he was uncertain about the causes and the evolutionary significance of these differences. His cousin Francis Galton was less circumspect, confidently drawing up a hierarchy of races in *Hereditary Genius* (1869) and arguing for intervention to speed up the development of the 'higher' races; he also wanted to encourage the extinction of the 'lower' races, such as Negroes, presenting this as a 'merciful form' of eugenics.[6] As eugenics gained momentum in Britain, the principal focus was on strengthening the home population and managing the feckless 'social residuum' associated with poverty and unemployment, and issues of racial difference were less prominent. After the Second World War, however, the question of race took on new significance as a result of Commonwealth immigration between 1948 and 1962. Reform eugenicists like C. P. Blacker were wary of becoming involved in this issue, recognising that of all the concerns of post-war eugenicists, race would be the most difficult to discuss publicly after the atrocities of Nazi genocide. Nonetheless, G. C. L. Bertram, his successor as secretary of the Eugenics Society, actively campaigned on the question of West Indian immigration, and eugenic views on race were also expounded by a number of well-known biologists and psychologists, most notably Cyril Darlington and Hans Eysenck. In their journalism and popular science books, these writers emphasised the biological basis of race and represented racial 'others' as a threat to the health of the body politic.

This chapter explores the relationship between eugenics and race at mid-century, at a time of transition when biological concepts of race were first being seriously challenged by the idea of race as a social construction. The 'Statements on Race' issued by the United Nations Educational, Scientific and Cultural Organization (UNESCO) in 1950 and 1951 offer a rich case history of this debate and of its wider social and political contexts. In the immediate aftermath of the Second World War, UNESCO was charged by the United Nations with disseminating 'scientific facts designed to bring about the disappearance of that which is commonly called race prejudice'.[7] The leaders of the project were convinced that anthropology and the social sciences had developed to the point where an authoritative rebuttal of the race science of the nineteenth and early-twentieth centuries was possible.[8] Accordingly, the 1950 statement declared firmly that '[f]or all practical social purposes "race" is not so much a biological phenomenon as a social myth'. It stressed that there was no evidence for differences in 'intelligence and temperament' between races, and it rejected the idea that there was any biological justification for preventing marriage between races (3, 8). Somewhat paradoxically, however, it endorsed the idea that there were three 'major divisions' in humanity, Mongoloid, Negroid and Caucasoid. The statement received a great deal of positive press coverage but provoked a storm of criticism from the scientific community, with the Royal Anthropological Institute, for example, arguing that the claim that 'race is less a biological fact than a social myth' was too simplified, and that there was no scientific evidence for the idea, also put forward in the statement, that human beings had an innate drive towards brotherhood.[9] There was a widespread perception that the statement had ignored the work of geneticists with specific expertise in this area and had been unduly influenced by social rather than physical anthropology. In response to scientific and public pressure, the project leaders convened a new panel with more input from geneticists and physical anthropologists. The second statement, which was published in 1951 along with a lengthy commentary, represents a retreat from the progressive optimism of the first but reflects more accurately the state of scientific thinking on race at the time. It distinguishes between the ethical principle of racial equality, which was unanimously supported by the contributors, and the concept of biologically based racial differences. The majority of participants thought that such differences did exist and endorsed the idea of race as a natural category. The rapporteur Leslie Dunn summarised the consensus view in these terms:

> The physical anthropologist and the man in the street both know that races exist; the former, from the scientifically recognisable and measurable congeries of traits which he uses in classifying the varieties of man; the latter from the immediate evidence of his senses when he sees an African, a European, an Asiatic and an American Indian together.[10]

Despite the fact that Dunn's appeal here is to the authority of physical anthropology (which coincides tellingly with the view of 'the man in the street'), it was the geneticists who provided the most detailed arguments in support of the 'reality' of the category of race. They argued that populations which were geographically separate would develop in biologically distinct ways, through the long term effects of genetic mutation and natural selection. In consequence, races would differ in a number of respects, including their level of intelligence. This issue came to dominate the 1951 statement. Whereas the first statement contained just one paragraph on race and intelligence, and concluded that there was no proof that races differed in their innate mental characteristics, the second devoted five paragraphs to a discussion of IQ tests, which, it was claimed, had shown differences in ability between different racial groups and between 'non-literate' and 'more civilized' peoples.[11] The discussion in the statement is confusing and often contradictory, so that within a single paragraph, we are told first that it is impossible 'to separate members of two groups on the basis of mental capacity', then that some types of 'innate intellectual capacity' may be more common in some racial groups (3). The confusion stems from the fact that there was little evidence that could be cited either to support or refute the claim for differences in intelligence. The debate thus turned on extra-scientific assumptions and beliefs, and it often involved argument by analogy rather than on the basis of evidence, as for example when Hermann Muller argued that it was likely that mental characteristics were subject to the same laws of heredity as other biological characteristics (53).

Muller was one of around 60 distinguished scientists who were invited to comment on the draft of the 1951 statement. Another was Cyril Darlington, who became a key figure in the development of the eugenic discourse on race in post-war Britain. This was in large part because Darlington's scientific work on chromosomes was being overtaken by advances in molecular genetics, which left him free to develop his interest in 'social genetics'. He was aware that many social scientists were suspicious of attempts to explain human behaviour in biological terms, but he wanted to persuade them of their mistake and to re-introduce biology into the study of man and society. Specifically, he wanted to intervene in the debate over the relative importance of heredity and environment and to make the case that human behaviour and the structure of societies, including the divisions of class and race, were genetically determined. He argued that human beings owed their rapid evolution to a unique combination of natural and 'artificial' selection. Natural selection ensured that in response to diverse environmental conditions, the qualities best adapted to survival in a specific environment were selected, while artificial selection (by which Darlington meant the complex human systems of caste and marriage taboos) accelerated the development of distinct, specialised races. Such specialisation, he suggested, had an important evolutionary function, preserving variety among the human gene pool. More importantly, however, the combination of specialisation and co-operation between races enabled the development of the complex societies that were characteristic of the human species. These ideas are explored at length in a trilogy of books directed at the general reader: *The Facts of Life* (1953), re-published in Pelican paperback as *Genetics and Man* (1964), *The Evolution of Man and Society* (1969) and *The Little Universe of Man* (1978).

Darlington's arguments were speculative and tendentious, deliberately so in the view of his biographer Oren Solomon Harmon, who suggests that Darlington set out to challenge what he saw as a conspiracy of silence among scientists over the question of race (250). Whether this was the case, his arguments are unsupported by genetic evidence, and reduce the complexities of social, political and economic history to a one-dimensional narrative of biological determinism. For example, in *The Facts of Life*, he develops his theory of racial specialisation and co-operation with specific reference to the racial composition of the US. He argues that the displacement of the Native Americans and the subsequent development of a 'white-plus-negro' society are genetically determined events, and that they have created a 'strong' society made up of genetically distinct races that nonetheless co-operate with each other. In his account, the Native Americans are presented as specialised, but lacking the capacity for social co-operation that would enable them to play their part in this society. As he puts it:

> The advantage of the white-plus-negro society is that its two racially dissimilar elements are able and willing (although only just willing) to help one another. The American Indian has been found not to be able and willing to help the other two. Their capacity and his capacity are both, of course, racial and genetic. They are determined and limited by heredity.[12]

Darlington's arguments about 'hybridization', the term he used for what eugenicists of the 1920s and 1930s had called 'race crossing', were equally tendentious.[13] His view was that, in some circumstances, 'outbreeding' (his term for miscegenation) was beneficial for human beings as it introduced variation into the gene pool. However, it could also be maladaptive, creating traits that were not suited to a particular race in a particular context. While stressing that hybridization 'is not in itself good or bad', his discussion of miscegenation in *The Facts of Life* emphasises the genetic damage it can cause:

> Following inbreeding a certain degree of purity or homozygosity is found in any old-fashioned group. . . . This pure-breeding is lost by crossing between races and castes and even classes. *The result of a single cross can be undone only by many generations of later inbreeding and selection.* (294, my emphasis)

Similarly, in an article published in 1950, he warns that 'random' human reproduction could lead to the permanent loss of valuable genetic material, and he raises the spectre of the untold damage that could be done through just one generation of 'outbreeding'.[14] Moreover, while he had been sceptical about the claims of eugenics earlier in his career, he was now convinced of the need for eugenic intervention to prevent such damage. Like Muller and Huxley, he was concerned that backward types were now multiplying out of all proportion to the more intelligent. Darlington's arguments in favour of eugenic measures were distinctive, however, in focusing on the implications of 'dysgenic pressures' with reference to the contract between the individual and the state. His argument was that in the post-war world it was taken as read that 'the State—and even the World State—should accept responsibility for the nutrition, health, and education of all human beings from the time of conception to the end of life'. However, this principle had a corollary: if the state took on this responsibility, it should have the right to make eugenic decisions in the overall interests of the population. This reasoning leads to an apparent endorsement of state eugenics when he argues that '[t]he State or mankind cannot accept a responsibility which goes back as far as the fertilised egg without one day claiming the right to go further and control the quality or proportions of gametes which go to make the fertilised egg'.[15]

Darlington published a number of articles in the *Eugenics Review* in the 1950s and 1960s and became more closely associated with the Eugenics Society. His perspective was very much in tune with that of the new secretary, Colin Bertram, who was a very different personality from C. P. Blacker. Blacker was the son of a diplomat who had himself displayed considerable diplomatic skills in modernising the Eugenics Society in the 1930s and in promoting its aims and agenda to the post-war Labour government.[16] But while Blacker remained influential behind the scenes, it was the more combative Bertram who led the eugenics movement forward in the post-war period, and he gained considerable publicity for it through his more direct approach. His broadsheet on West Indian immigration is a good example of this, as a very public statement of eugenic opposition to post-war 'coloured' immigration. Commonwealth immigration increased rapidly in this period because, under the terms of the 1948 British Nationality Act, all Commonwealth citizens were able to enter the UK freely, find work and settle with their families. The Act was primarily aimed at maintaining close ties with the white dominions, and Atlee's government was initially taken aback at the numbers coming from the Caribbean, in particular. However, it was soon seen that the immigrants could be extremely useful given a shortage of unskilled labour, and between the census dates of 1951 and 1961, the Caribbean population in the UK rose from 15,300 to 171,800; the Indian, Pakistani and 'Far Eastern' populations also increased rapidly in this period.[17] By 1962, however, there was popular and political agitation for tighter immigration controls, and the rights of Commonwealth immigrants were gradually whittled away through a series of measures over the next ten years, starting with the 1962 Commonwealth Immigrants Act.

The Eugenics Society chose to publish Bertram's *West Indian Immigration* broadsheet in October 1958, just two months after the Notting Hill race riots, which were sparked off by an attack on a mixed race couple by a gang of white youths. It is likely that Bertram wished to capitalise on the riots, thinking they would offer support for his arguments. The broadsheet was sent to a wide range of influential individuals, including all Members of Parliament, privy councillors and bishops, and it received extensive press coverage, with articles in the *Times*, the *Daily Telegraph* and other mainstream newspapers. In the pamphlet, Bertram follows Darlington's argument that the biological differences between races had come about because of geographical isolation and adaptation to different environments. He goes on to claim that 'the differing genetic endowments of people from different races' are apparent as early as the foetal stage of life, and that human races are anatomically almost as distinct as 'the breeds of domestic dogs'.[18] In a sleight-of-hand characteristic of eugenic discourse, he disclaims any intention of making 'judgements of relative worth' before going on to raise the value-laden concept of intelligence and reflecting that it would be remarkable if racial differences in intelligence did not exist. His main concern, however, is with miscegenation, which, he argues, 'runs counter to the great developing pattern of human evolution', because it works against the level of specialization that Darlington, among others, thought so important for the advancement of mankind (19). For this reason, Bertram argues that 'cross breeding' is more acceptable between individuals of genetically close races than between those that are 'markedly different'. In general, he claims, there is a widespread distaste for 'cross breeding': this is likely to spring from 'inborn factors' and may serve an evolutionary purpose, by preventing race mixture. This response is described in markedly somatic, affective terms: it is a 'colour feeling or resistance' that provokes strong emotions in both individuals and 'the wider groups to which they belong' (18). In a book published a year later, *Adam's Breed*, Bertram states more explicitly that this distaste serves a biological purpose in protecting the distinctiveness of races. Again, he emphasises the somatic and sensory dimensions of this reaction, using images of smell and taste:

> But when my neighbour and myself are of different colour, different smell and different shape—genetically quite a lot, not just a little,

> different—and our food habits, marital habits and traditions are greatly at variance, then there is real potential antagonism between us.[19]

The physical imagery works to construct a sense of a substantive disjunction between the white and the black races; the suggestion is of a biological incompatibility that makes 'race mixture' into a form of pathology (Figure 4.1).

Figure 4.1 Maj-Britt Morrison leaves the West London Police Court with her husband Raymond having been charged with obstruction during the Notting Hill race riots, 1 September 1958.

Such constructions of difference are central to all racial discourse, but what is distinctive about post-war eugenics is the attempt to legitimate the biological concept of race by an appeal to the authority of genetic science. Neither Bertram nor Darlington had any hard evidence to support the claims they made, but they were able to draw on aspects of Huxley's Modern Synthesis to justify what were only speculations about the existence of biological racial distinctions.[20] In particular, they were able to mobilise genetic explanations of racial difference to underwrite their fixed opposition to miscegenation. As numerous scholars have shown, miscegenation is at the heart of racial discourse, the site on which both fear and desire of 'the other' are projected and inscribed. In the context of eugenics and race, it is also understood as the key site on which the integrity of the body politic is breached and the strength of the population weakened. The extent to which such attitudes filtered through into popular consciousness is illustrated in a survey of attitudes in universities in the 1950s. A hundred female students at Oxford and Cambridge were asked to fill in a 'social distance card' in relation to race, and in answer to a specific question, 84 said they would not be willing to marry an African (no distinction was made in the question between different parts of Africa). According to the researcher, '[m]any of those who said they were unwilling to do so justified their views by reference to "the misery of half-caste babies", to Africans' "lack of intelligence" or to the undesirability of "mongrelisation"'.[21] The use of the term 'mongrelisation' echoes Bertram's comparison between human races and different breeds of dog, and it suggests an endorsement of the idea of racial 'purity'.

George Lamming's novel *The Emigrants* (1954) offers a trenchant response to such conceptualisations of race mixture in terms of pathology and the threat of degeneration. In a number of respects, Lamming takes racial discourse and turns it back on itself, as, for example, when he uses the trope of vomit to figure the multiple exchanges that have formed the population of the Caribbean. After the indigenous Caribbeans were wiped out, one of his Jamaican protagonists explains:

> England, France, Spain all o' them, make them vomit up what them din't want, an' the vomit settle there in that Caribbean Sea. It mix up with the vomit them make Africa vomit, an' the vomit them make India vomit, an' China an' nearly every race under de sun. . . . Them settle right there in that Caribbean sea, and the great nations, England, an' the rest, them went on stirring the mixture, them stir that vomit to suit themself.[22]

Lamming here draws attention to the history of forced migration to the Caribbean, which he represents as the dumping ground for 'rejects' from within the colonial countries, as well as those traded as slaves from Africa, India and China. In using the metaphor of vomit, he figures such people in terms of sickness or contagion, something that had to be expelled from the 'home' body politic to preserve its health. The subsequent 'mixing' of the people is described, however, in transformative terms. The Jamaican suggests that as the great nations 'stir' the vomit, it begins to take on its own life:

> it was like ammonia, get too strong for those who stirrin' it. Now it explodin' bit by bit. It beginnin' gradjally to stir itself, an' you can understan' what happenin' if you imagine yuh vomit take on life an' start to find out where yuh stomach is . . . that vomit feel funny, queer, where it settle. It want to find a stomach 'cause it realize that it is expose'. It stirrin' itself but there ain't no pot. (67)

Here 'race mixture' is represented in terms of an evolutionary advance. The 'mixing' has created a new organism, which is both vulnerable (exposed) and potentially extremely powerful (explosive). In this account, the biological material that was previously rejected has turned into a force to be reckoned with, not least by the colonial powers, who will be confronted in the future by 'a great people', anxious to prove that they 'be something'—and also anxious to find a new 'pot', in other words a new home.

The Emigrants also explores the theme of miscegenation at the individual level through a narrative thread involving Philip, a middle-class boy from Trinidad who is studying to be a lawyer, and his girlfriend Julie. When Julie becomes pregnant, her husband threatens to divorce her, but if this happens, Philip will be cited as the co-respondent and will lose his scholarship. He discusses the possibility of an abortion with his friend Azi, who organises it in such a hurry that it has already happened by the time Philip decides that this is not what *he* wants. In his treatment of the theme of miscegenation, Lamming again reverses conventional racial discourse. In *The Emigrants*, the abortion is not the result of a white family preventing the birth of a mixed race child on quasi-biological grounds. Instead, an African, Azi, acts in what he thinks are the best interests of his Caribbean friend: he places Philip's ambitions above the needs of Julie and the child, as he is convinced that a mixed race marriage would put an end to Philip's career. The whole episode is one of reversals and unnerving displacements, as when Julie has her abortion upstairs in a sleazy club, and the violence enacted on her body is refracted through an attack on another woman, whose blood streams out on the dance-hall floor: '[t]he woman howled like a sick animal . . . [she] had made a pool of blood on the floor' (270).

RACE RELATIONS

The discourse of 'race relations' was developed in opposition to eugenic and genetic racism, its premise being, as with the first UNESCO statement, that race was not a biological phenomenon but a social construction. It was pioneered by anthropologists and sociologists such as Kenneth Little, Sheila Patterson and Michael Banton. Little had started out in the 1940s as a physical anthropologist measuring the heads of black children in Cardiff, but his research led him to the inescapable conclusion that it was social factors (such as poverty, poor housing and lack of education) that created so-called racial differences. He switched to social anthropology and with his students developed a number of interpretive models for the analysis of relationships between races, including the concept of the host/stranger relationship and of interactions between 'in groups' and 'out groups'. The work of these of researchers was influential in both academic and popular circles, appearing in major academic journals such as the *Sociological Review* and the *British Journal of Sociology* and finding a wider readership in Pelican paperbacks such as Anthony H. Richmond's *The Colour Problem* (1955) and Sheila Patterson's *Dark Strangers* (1965). However, as Chris Waters points out, their findings must be approached with caution, as they were not always aware of the extent to which their own assumptions shaped the outcomes of their enquiries and constructed the 'migrant experience' they sought to document.[23] Moreover, while they denied the biological basis of race, they assumed the existence of 'naturally' formed social groups that were organised around differences in 'colour'. Biological concepts of race were thus translated into social terms but at the same time fixed and reified. For example, in Sheila Patterson's analysis of the differences between West Indians and the white community in Brixton, the popular belief in the 'natural' promiscuity and fertility of black women is re-inscribed in terms of cultural belief systems when she writes that for West Indians '[s]exual intercourse is regarded as a natural activity . . . [f]ecundity in a woman is seen as natural and desirable, while barrenness and artificial prevention are thought of as unnatural, unhealthy, even wrong'.[24] In the work of Michael Banton, biological and sociological explanations of difference are also ambiguously intertwined. In his account of relationships between white women and black men, he argues that the majority of women in these relationships come from 'a background of deprivation' that has made them 'psychologically abnormal', thereby explaining their behaviour in terms of a socially constructed pathology. However, he then goes on to invoke notions of natural difference when he suggests that many of the women are attracted to the physical strength of their black partners because they 'derive pleasure from physical punishment'.[25] Similarly, in analysing the causes of the widespread prejudice against miscegenation, Banton struggles to separate social and biological discourses of race, arguing that:

> in view of the emphasis in English culture upon 'breeding true' and upon the importance of a nation's maintaining its cultural identity, it is perfectly natural for people who have been brought up in these ideas to object to something which will produce what they regard as 'mongrels'—offspring who do not conform to the preferred cultural pattern. (136)

Here the discourse of social construction is destabilised by the use of terms drawn straight from the lexicon of eugenics; moreover, when Banton writes of the failure of popular opinion to distinguish between social and biological factors in race, he clearly underwrites the idea that *both* do in fact exist.

The equivocations that mark the discourse of race relations are also evident in texts by white British writers that reflect on race in this period. The journalist and novelist Colin MacInnes, for example, has been praised for his engagement with marginalised communities, and the essays he published in *New Society* in the 1960s helped to draw attention to the new identities and relationships that were emerging as a result of post-war immigration.[26] Yet the sexual and racial politics of his novel *City of Spades* (1957) are ambivalent and frequently disquieting. The text has a split structure, juxtaposing the first person narratives of Montgomery Pew, an immigration welfare officer, and Johnny Fortune, a Nigerian immigrant. Pew's perspective is ostensibly that of an 'enlightened' white man schooled in the discourse of race relations, yet his narrative is shot through with a homoeroticism in which the black male is sexualised and objectified. For example, at one point, Pew describes 'a huge and handsome African, who positively dripped and oozed with mindless masculine animal magnetism'.[27] The imagery is suggestive of a creature who has just emerged from the primeval slime, and for Pew, the attractiveness of the black male body seems to depend on the fact that he can represent this body to himself as 'mindless'. On another occasion, Pew muses on the 'primitivism' of a troupe of black dancers in terms that recall D. H. Lawrence's investment in the notion of the savage possessed of prehistoric and pre-linguistic knowledge:

> as they danced, they were clothed in what seemed the antique innocence and wisdom of humanity before the Fall—the ancient, simple splendour of the millennially distant days before thought began, and civilisations . . . before the glories of conscious creation, and the horrors of conscious debasement, came into the world! In the theatre, they were *savages* again: but the savage is no barbarian—he is an entire man of a complete, forgotten world, intense and mindless, for which we, with all our conquests, must feel a disturbing, deep nostalgia. (177, MacInness's emphasis)

The use of this trope by MacInnes and other writers of this period looks back to modernism but also anticipates, and perhaps feeds into, the controversy over race and intelligence which broke out in the late 1960s and 1970s.[28]

MacInnes also pathologises miscegenation, representing it in terms of a repeating pattern leading to degeneration. Johnny Fortune's father has had a child with a white woman, Mrs Macpherson; Johnny, in turn, has a child with her daughter Muriel. Mrs Macpherson is characterised as bitter and vengeful, having been left to rear 'a bastard nigger child' on her own, and her mixed race son conforms to a number of eugenic stereotypes about the 'half-caste'. He is mentally limited ('grinning with imbecile guile'), has 'his mother's crazy glance' and is trapped in a cycle of criminal activity interspersed with spells of imprisonment (83, 88). The prospects for Muriel's son with Johnny look a little brighter, but this is largely because of the intervention of Johnny's white friends, who offer financial and other support to Muriel. Johnny, meanwhile, goes back to Lagos, conforming to the pattern described by social scientists like Patterson of the feckless black father who takes no responsibility for his children.[29] Like the 'race relations' scientists, MacInnes thus re-inscribes racial stereotypes in his attempts to engage on equal terms with black 'others'.

Muriel Spark's short story 'The Black Madonna', on the other hand, offers an incisive critique of the self-deception that can underlie 'enlightened' attitudes to race like those promoted by the race relations theorists. The story is located in a new town near Liverpool, a port with a long-established black community. It focuses on Lou and Ray, a childless Irish Catholic couple who have distanced themselves from their working-class background and aspire to be liberal intellectuals, reading the left-wing *Observer* newspaper and looking down on their neighbours' addiction to mass culture. When two Jamaicans, Henry and Oxford, come to work in Ray's department at the local motor works, he takes them under his wing, and Lou introduces them to all her friends, whom she instructs in the correct way to deal with black people ('don't call them darkies').[30] However, when Henry criticises her sister, who lives in a dirty East End slum, Lou's underlying racism comes to the surface:

> 'That sort,' said Henry, leaning forward from the back of the car, 'never moves. It's the slum mentality, man. Take some folks I've seen back home.'
>
> 'There's no comparison,' Lou snapped suddenly, 'this is quite a different case.'
>
> Raymond glanced at her in surprise; Henry sat back, offended. Lou was thinking wildly, what a cheek *him* talking like a snob. At least Elizabeth's white. (21)

After this incident, the couple and the Jamaicans drift apart, an outcome Lou ascribes to the prayers she has offered on the subject to the Black Madonna, an icon that has recently been installed in the local church. The Black Madonna can also, it is alleged, help infertile couples, and encouraged by her success with the Jamaicans, Lou decides to 'put in for a baby' to her. She duly becomes pregnant—and gives birth to a black baby. She refuses to believe that the baby is hers, and has to be heavily sedated after 'an attack of screaming hysteria' (28). However, tests are carried out, and it transpires that the baby is indubitably hers and Ray's, with the 'black blood' coming from Lou's side of the family, as part of her 'obscure Liverpool antecedents' (29). Lou insists that the baby be put up for adoption and tells Ray that the priest thinks they are doing 'the right thing'. As Spark is at pains to emphasise, the right thing is not necessarily a good thing, and from a Catholic perspective, the couple are placing expediency before virtue. Yet the story makes a point that goes beyond the finer points of theology, as Spark exposes a post-war British culture that reads a deep and fundamental difference into visible, so-called 'racial' characteristics. For Lou and Ray, the visible signifier has far more meaning than their (verified) biological relationship with their daughter, so that her colour effectively divorces her from her parents. Spark here exposes the formidable power of what Fanon calls epidermal racism, which in this instance overrides the supposed 'truth' of genetic inheritance.[31]

INVASION OF THE BODY POLITIC

There was a hardening of attitudes to race in Britain in the 1960s, which seems out of step with the socially progressive attitudes of the time but can be linked to economic instability and its impact on a particular employment sector. As Anna-Marie Smith has argued, new pressures on the British economy 'accelerated the disruption of white workers' positions in the manufacturing sector and the disintegration of white working-class neighbourhoods'.[32] The majority of black workers were in unskilled jobs and lived in very poor conditions, but as white workers' jobs were threatened, the lines of demarcation between the white working-class and black communities threatened to dissolve. In this context, an anti-black immigration movement took root and expanded rapidly. It included groups such as the Birmingham Immigration Control Association and the Union Movement, which incorporated elements of Oswald Mosley's pre-war British Union of Fascists. A number of Conservative MPs began to lobby for immigration controls, and in 1964, Peter Griffiths won the Smethwick seat in Birmingham with an explicit anti-black immigration campaign, in which he spoke of the need to stop 'criminal' and 'unhealthy' immigrants from entering the country. As Smith points out, popular feeling against immigration was also fuelled by racist interpretations of two specific events. In 1967, there was a small influx of Kenyan Asians who came to the UK when the Kenyan government removed their citizenship rights.[33] Their numbers were small, but Enoch Powell and other right-wing politicians represented this as a significant immigration 'crisis'. In addition, on 4 April 1968, Martin Luther King was assassinated in the USA, and the demonstrations and riots that followed received wide press coverage, accompanied by commentary in the right-wing press suggesting that similar violence could erupt in Britain.

It was against this backdrop that Enoch Powell made his infamous 'rivers of blood' speech at the Birmingham Conservative Political Centre on 20 April 1968. Powell was a skilled rhetorician, and his speech was carefully crafted, but although its articulation of the intensified racism of the period has been analysed in great detail, less attention has been paid to its eugenic dimensions. The speech is organised around two anecdotes, purporting to illustrate the views of 'ordinary, decent, sensible people'. In the first, Powell recounts a conversation with one of his constituents in Wolverhampton, who expresses his fears about the future of his children and grandchildren:

> He suddenly said: 'If I had the money to go, I wouldn't stay in this country . . . I have three children, all of them been through grammar school and two of them married now, with family. I shan't be satisfied till I have seen them all settled overseas. In this country in 15 or 20 years' time the black man will have the whip hand over the white man'.[34]

As many commentators have noted, in using the metaphor of the 'whip hand', Powell alludes to and inverts the powerful imagery of slavery and the power of the white plantation owner, pointing to a potential reversal of authority. However, in foregrounding his constituent's fears for his children and grandchildren, he is also drawing attention to the ties of blood and kinship, and to the paradox whereby the future of the man's family can only be secured through fracture and dispersal, in other words, emigration. Having conjured up this image of a fractured family, and by implication of a fractured (white) body politic, Powell goes on to make his eugenic point plain. Britain, he claims, is committing a kind of race suicide, inviting obliteration by the invading 'others':

> We must be mad, literally mad, as a nation to be permitting the annual inflow of some 50,000 dependants, who are for the most part the material of the future growth of the immigrant-descended population. *It is like watching a nation busily engaged in heaping up its own funeral pyre.* So insane are we that we actually permit unmarried persons to immigrate for the purpose of founding a family with spouses and fiancés whom they have never seen. (my emphasis)

Powell portrays the arrival of the dependants as having the potential to swamp the indigenous population, as their sole purpose in coming to Britain is, it seems, to have families.

The emphasis on the reproductive power of the immigrants is extended into the second of Powell's anecdotes, which juxtaposes the ageing, post-reproductive body of a white female pensioner with the vitality of the black population. He quotes from a letter sent to him by a concerned constituent:

> Eight years ago in a respectable street in Wolverhampton a house was sold to a Negro. Now only one white (a woman old-age pensioner) lives there. This is her story. She lost her husband and both her sons in the war. So she turned her seven-roomed house, her only asset, into a boarding house. She worked hard and did well, paid off her mortgage and began to put something by for her old age. Then the immigrants moved in. With growing fear, she saw one house after another taken over. The quiet street became a place of noise and confusion.
>
> . . . She is becoming afraid to go out. Windows are broken. She finds excreta pushed through her letterbox. When she goes to the shops, she is

> followed by children, charming, wide-grinning piccaninnies. They cannot speak English, but one word they know. 'Racialist', they chant.

The fact that the un-named woman has lost her husband and two sons in the war symbolises the losses that eugenicists saw as a real threat to the British population in the post-war years. Now bereft of her husband, children and any prospect of grandchildren, she is almost literally under siege from the invading 'others' and their taunting children. Even the internal space of her home is no longer secure: its boundaries are breached as the windows are smashed in and excrement is pushed through the letterbox.

The trope of excrement links immigration with infection and contagion, and Powell's deployment of such metaphors points to the eugenic concerns that underlie his racist rhetoric. For him, black immigration represents, as he says, 'an invasion of the body politic', and he is concerned not just with the rate of immigration, but with the allegedly high birth rate among the immigrants. Powell's rhetoric was particularly significant in this respect, as it created a novel framework for the perennial eugenic concern with differential breeding. So whereas in the inter-war period eugenicists had focused on the threat posed by the British urban poor, they now turned their attention to the dangers associated with a disproportionately high birth-rate among the immigrant population. Powell goes into great detail about this, citing statistics about relative birth-rates in his Wolverhampton constituency and elsewhere. What particularly concerns him is that this expansion is taking place in what he calls 'encampments', in other words, areas in which the immigrant population has become densely concentrated. This is significant from a eugenic point of view because, he argues, racial minorities will only achieve a normal birth-rate when they have been 'thoroughly absorbed into a host population'.[35]

Fugue for a Darkening Island (1972), the acclaimed first novel of the science fiction writer Christopher Priest, offers a complex response to Powell's rhetoric and sheds light on the eugenic anxieties that fuelled anti-immigration discourse. The text is mediated through the consciousness of the first person narrator, Alan Whitman, an academic who (as his name suggests) acts as a representative of the ostensibly liberal, white middle-class. The novel is set in the near future, at a time when the racial violence predicted by Powell has developed into armed conflict. The premise is that a short but catastrophic nuclear war in Africa has left millions sick and starving and has led to an exodus across the globe. When the first refugees (the 'Afrims') arrive in London, Whitman is one of the thousands who watch them landing with 'a mixture of horror and fascination'.[36] Most of the passengers on board are in an advanced stage of starvation, with skeletal arms and legs; they are naked, and many of them cannot walk or are lying unconscious; the police later find more than seven hundred corpses on the ship, most of them children. In this haunting scene, Priest creates a layered image that condenses the arrival of these refugees, memories of Commonwealth immigrants arriving in Britain on ships such as the SS *Empire Windrush*,

(Figure 4.2) and the slave ships of the middle passage. Like Lamming, he thus locates post-war immigration in the wider context of Britain's colonial history and its involvement in the slave trade—a context that was elided in most post-war debates about immigration.

Figure 4.2 The SS *Empire Windrush* arrives at Tilbury Docks from Jamaica, 22 June 1948.

At the point when the Afrims arrive, Britain has a neo-racist government led by John Tregarth, a figure modelled on Enoch Powell, who espouses policies of 'patriotism, nationalism and racial purity'. Tregarth's response to the crisis is one that Powell would have endorsed, as a new Order Act is passed in order to restrict 'the rights and movements of the aliens' (110). Government hostility to the refugees escalates the conflict, and eventually Britain is caught in a three-way civil war between the Afrims and the Nationalists, with the Secessionists, a party of white supporters of the Afrims, placed somewhere in the middle. As a liberal intellectual, Whitman is ostensibly aligned with the Secessionists, but his commitment is little more than a matter of form. On a conscious level his actions are driven by a mixture of expediency and inertia, while on a subconscious level he is in the grip of inchoate but deeply ingrained eugenic prejudices. When the unrest begins, his instinct is to avoid taking sides, mainly because he is preoccupied with a love affair. Eventually he joins a college society that is allegedly 'sympathetic to the plight of the Africans', but there is no real commitment among the members, and he himself has joined only because he fears being isolated at work. As the situation worsens, he and his family have to leave London and fend for themselves in the countryside, and eventually his wife and daughter are abducted by the Afrims. This abduction confirms the feelings of sexual and reproductive failure that pervade his thinking and that he associates with a more general loss of vitality among the white population.

The novel's structure is fragmented and a-chronological, and the blanks in the text contribute to the sense that Whitman is on the edge of breakdown (echoing the idea of the fragmentation of the individual and the body politic in Powell's 'rivers of blood' speech). At the heart of this fracturing of identity are feelings of sexual inadequacy. Whitman's marriage is sexually dysfunctional (he and his wife have not had sex since the birth of their only daughter), and his love affairs are affect-less and give him little pleasure. Moreover, he is haunted by fear of the superior sexual power of the black male, as is demonstrated by an episode in which he and one of his lovers watch a film about inter-racial sex. In this film, the affair between a coloured man and a white woman is 'violently-resolved', and as a result of watching it, Whitman is 'rendered impotent' and unable to have sex despite his girlfriend's 'best efforts' (19). Similar scenes occur in a number of novels of the period, but what is distinctive here is that the sexual humiliation of the white man is not associated with fears of miscegenation. Priest's apocalyptic vision goes further, mapping out scenes of genocide in which the white female reproductive body is the locus of black violence. In one scene, for example, Whitman discovers the bodies of four young white women:

> Each of the women was young and had been physically attractive. Their deaths had come after a long period of helpless agony: the torment was embedded in their expressions. Each one was tied hand and

> foot, and had evidently struggled to escape from the bonds in her last few minutes of life.
>
> The men who assaulted them had disfigured their bodies with either bayonets or knives, slashing them many times in the region of their genitals. There was blood all over the floor. (97)

The repeated slashing of the women's genitals signals the eugenic dimension of these assaults, which aim to destroy the reproductive power of the white population.

In depicting a future in which the Afrims are effectively engaged in ethnic cleansing, Priest offers a mirror image of the eugenic fantasies that underpinned Powell's rhetoric and highlights the mobility and reversibility of eugenic logic. This point is underscored by the final twist in the novel, when Whitman discovers the bodies of his lost wife and daughter, who have been dumped on a beach after being tarred with black pitch. Impelled by a 'disturbing combination of terror and hatred', Whitman takes off with a rifle, intent on killing every African he can find (147). This is a science fiction text, and Whitman's trajectory is extreme, but the underlying implication is that the seeds of such behaviour are already present among the apparently docile British middle-classes. This is the distinctive achievement of Priest's novel, to suggest the extent to which eugenic anxieties haunted the imagination of the educated middle-class.

RACE AND INTELLIGENCE

The alleged link between race and IQ was at the heart of eugenic concerns about race, and in 1969, the long-running debate on this issue was re-ignited with the publication of an article by Arthur R. Jensen in the *Harvard Educational Review*, 'How Much Can We Boost IQ and Scholastic Achievement?'[37] Jensen, an educational psychologist, had carried out a review of the extensive literature on heredity and intelligence that had been built up in the US since the 1920s and concluded that in IQ tests blacks scored on average about 15 points below whites. Taking issue with what he called the 'zeitgeist of environmentalist egalitarianism', he insisted that genetic factors were implicated in these differences. The article received a great deal of media coverage and provoked heated debate on both sides of the Atlantic. By far the most devastating critique of Jensen's work came from the Harvard geneticist Richard Lewontin, who argued that his ideas were based on a fundamental conceptual error, because heritability estimates can be applied only to the analysis of differences *within* groups, whereas Jensen had employed heritability to help explain the difference in scores *between* two groups (i.e., blacks and whites). In consequence of this methodological confusion, he contended, Jensen's conclusions were 'meaningless'.[38] Lewontin's work has been of crucial importance in discrediting the idea that there is any genetic basis for 'racial' distinctions; for example, his statistical calculations have demonstrated that the genetic variation between populations classed as belonging to separate 'races' is around 6%, in other words, it is scientifically negligible. The significance of Lewontin's intervention in the Jensen debate was immediately recognised, and he was supported by a number of other distinguished geneticists who drew attention to the wide range of factors that were likely to produce apparent differences in intelligence, including the pre-natal environment and poor diet, as well as a range of cultural factors.[39] Undeterred, Jensen continued to promote his ideas, with the assistance of the physicist and well-known eugenicist William Shockley. In Britain, his arguments were enthusiastically supported by Hans Eysenck, who had supervised Jensen's early post-doctoral research. After participating in a public debate over Jensen's claims, which was organised in Cambridge by the British Society for Social Responsibility in Science, Eysenck published *The IQ Argument: Race, Intelligence and Education* (1971). In this short book, he summarised Jensen's arguments and added his own contentious views on 'scholastic backwardness' among black children in Britain. Eysenck was at this time the best-known psychologist in the UK. His fame as a public intellectual ensured widespread press coverage for his book, and his position as head of the prestigious Maudsley Institute of Psychiatry lent credibility to his arguments. After the book's publication, he was hailed as a hero by the political right and by the neo-Nazi National Front movement, while on the other side of the argument, radical student groups attacked him and flour-bombed his lectures. The situation was sufficiently serious for the National Union of Teachers to take the unusual step of publishing a pamphlet that was distributed to every one of its members, countering Jensen's and Eysenck's arguments and explaining their methodological flaws.[40]

The IQ Argument reflects an important shift in the longstanding concern with IQ among British eugenicists. Like Colin Bertram, who in 1958 had advocated IQ testing for all immigrants entering the UK, Eysenck focuses attention on the black population as a threat to the intelligence of the nation. In addition, he invokes the epistemological authority of genetic science to endorse his arguments, despite the fact that, like Jensen, he had no expertise in this area. In fact, although both Eysenck and Jensen speak of 'genetic' differences, their methodology comes from educational psychology and includes methods of statistical analysis, which were by then outdated. They also drew on data taken from the twin studies of Cyril Burt, which were soon to be proved fraudulent. On this fragile basis, Eysenck declares confidently that the ratio between genetic and environmental influences on intelligence is 4:1, in other words, that genetic factors account for 80% of observed differences in intelligence.[41] He goes on to cite a study from the US in which, even when white and black men were matched in terms of their social status and environment, there was a difference of 10 to 14 points in their IQ scores. A large number of other studies are referred to, including one which demonstrated that there were far fewer 'gifted' black children in US schools, although there were significantly more black children who fell into the category of the 'retarded'. In a particularly startling move, Eysenck then compares the (alleged) lower intelligence of black Americans with the condition of phenylketonuria, an inherited metabolic disorder that, if untreated, causes severe learning difficulties. Eysenck suggests that, just as treatment for this disorder came out of an understanding of its hereditary nature, so interventions aimed at boosting IQ in other groups are likely to depend on recognition of the hereditary causes of low intelligence.

Eysenck claims that in raising these issues he is motivated by the desire to establish the 'facts' about race and intelligence, and while he acknowledges that these 'facts' are not yet fully known, he nonetheless goes on to make some suggestions for improvement that involve, in effect, differential education for black children. Building on Jensen's contention that black children have a lower 'conceptual IQ' but a relatively high 'associative learning ability', he advocates methods of teaching specifically targeted at 'associative' skills. His recycling of theories about 'innate' differences that were developed in relation to black children in the US overlapped with discourses of cultural 'backwardness', which were then being mobilised in Britain to explain the poor school performance of black children. A report by the National Foundation for Educational Research in 1966 had concluded, for example, that immigrant children were 'handicapped' by a 'defective education' in their country of origin, by 'family instability' and by the low cultural level of their home environment.[42] A government report on *The Problems of Coloured School Leavers* (1969) therefore recommended 'special provision' for black schoolchildren, on the grounds that '[e]quality of opportunity does not always mean treating everyone in exactly the same way. . . . Special problems need special treatment' (192). As Hazel Carby argues, such policy statements signalled a transition from indirect to more direct control over the schooling of black children, which in turn led to a disproportionate number of black children being placed in the lowest streams in comprehensive schools or diagnosed as educationally sub-normal.

Although Jensen was not a geneticist, the idea that there were genetically based racial differences in intelligence was endorsed by many in the genetics community. The most eminent of his supporters was Francis Crick, then based at the Cavendish Laboratory in Cambridge, whose interest in eugenics has been discussed in the previous chapter. Crick expresses his eugenic views bluntly in a number of letters written to Peter Medawar in the 1970s. For example, he makes the suggestion that 'the bulk of people generally regarded as valuable to society [should] be encouraged to have more children and those who are a handicap [should] be encouraged to have less', as this 'would make for a much more attractive and acceptable set of people'. In a similar vein, he comments, 'I find it difficult to believe that the present gene pool is near *any* sort of optimum composition and rather surprising that no one dares to consider whether a different gene pool might not be more acceptable to almost everybody contained in it'. He also considers that 'such a policy in *the long run* (i.e. within the next 100 years) is virtually certain to be tried'.[43] The part that Crick played in the Jensen controversy thus needs to be placed in the context of his long term, sustained engagement with eugenic ideas.

Crick's interest in the race question is evident from the many papers he kept on the subject, which include offprints of articles sent to him by Jensen, the Lewontin paper on 'Race and Intelligence', which refuted Jensen's claims, photocopies of a number of relevant articles from *Science* and cuttings relating to the controversy in the British press. Crick corresponded with Jensen and corrected a statistical error in one of his papers, and in a letter to the Harvard sociologist John T. Edsell expressed his broad agreement with Jensen's views:

> Unlike you and your colleagues I have formed the opinion that there is much substance to Jensen's arguments. In brief I think it likely that more than half the difference between the average I.Q. of American whites and Negroes is due to genetic reasons, and will not be eliminated by any foreseeable change in the environment. Moreover I think the social consequences of this are likely to be rather serious unless steps are taken to recognize the situation.[44]

His support for Jensen was not confined to comments in private letters. After the publication of his article, Jensen was subjected to disruption of his lectures and classes, and at the height of the controversy, the *Harvard Educational Review* stopped selling the issue of the journal in which his essay was published. A number of scientists saw this as a threat to academic freedom, and Ellis B. Page, a professor of educational psychology at the University of Connecticut, drafted a 'Resolution on Scientific Freedom' in response. Jensen sent a copy of the draft to Crick, who contacted four eminent British biologists and asked them to join him in signing it. Sydney Brenner and John Kendrew agreed, but Frederick Sanger declined, on the grounds that, although he supported the general idea, he was not a specialist on questions of inheritance. Maz Perutz firmly refused, despite Crick's pressing him on the issue.[45] The 'Resolution', which was published in the *American Psychologist* in July 1972, was signed by Cyril Darlington, Raymond Cattell and Hans Eysenck, among others. As a statement, it can be read on a number of levels. On the face of it, it represents a defence of the right of scientists investigating the role of heredity in human behaviour to publish their findings without 'censure, punishment, and defamation'.[46] However, it also makes exaggerated and inflated claims, comparing Jensen's position with that of the Mendelian geneticists suppressed by Stalin, for example. In addition, it presents an extremely biased account of the differences of opinion between biologists and social scientists on this issue. The biologists are presented as weighty and authoritative figures; it is stated that they have 'investigated much evidence' and that their arguments are 'well-grounded'. On the other hand, the sociologists, social psychologists and social anthropologists are accused of evasiveness and lack of responsibility in constructing their arguments. What is supposed to be an intervention on a matter of principle descends into an attack on the credibility of the social sciences.

In Britain (as in the US), this controversy had damaging and long term effects. Jensen and Eysenck's claims about race and intelligence were vigorously contested not just by social scientists but by biologists such as Steven Rose, who wrote a stinging attack on a subsequent book by Jensen, *Educability and Group Differences*, in which he rehearsed the same arguments as in his 1969 article.[47] Despite such challenges, Jensen's claims played into and endorsed 'common sense' racist beliefs, and notions of innate difference became increasingly prominent in discussions about the poor performance of black schoolchildren. The idea of the black immigrant as potentially defective also made its way into the popular imagination, as the British-Nigerian novelist Buchi Emecheta testifies in her semi-autobiographical first novel *In the Ditch* (1988). Emecheta came to London in 1960 to join her husband. It was a difficult marriage, and in 1966 she left him to study for a degree in sociology while working to support her five children. *In the Ditch* has its origins in a regular column she wrote for the *New Statesman*, and it offers a sharp critique of the discourse of race relations and in particular of the pathologisation of the black family at this time. In relation to the question of intelligence, the narrator of *In the Ditch* comments, with respect to the main protagonist Adah, that '[o]ne of the methods she had found very helpful in securing friendship in England was to pretend to be stupid. You see, if you were black and stupid, you were conforming to what society expected of you'.[48] Emecheta also picks up on the way in which the eugenicist concept of the 'problem family' was reconfigured in the 1960s and 1970s to focus on the families of black immigrants. So as Adah drily notes, 'a coloured family sandwiched between two white ones' will frequently be defined as a problem family' (117). Such pathologising of black families paved the way for interventions aimed at bringing their behaviour in line with the desirable (white) norm.

RACE AND FAMILY PLANNING

Although West Indians were the primary focus for eugenic anxieties about race between 1945 and 1980, South East Asians were also a target by the late 1970s as a result of increased immigration from India and Pakistan. However, throughout the post-war period, the family structures of West Indians and Asians were constructed and interpreted in markedly different ways by sociologists and policy makers. In her study of Mirpuri Asians in Bradford, for example, Verity Khan argues that their households provide a structured and cohesive atmosphere in which each member of the family is 'conscious of her place and responsibility to others in the wider kinship group'; this system promotes 'a certain stability and psychological health in all individuals'.[49] Michael Banton makes a similar point about the shared values that sustain Indian and Pakistani communities, but he also emphasises their distance from the 'host' community, arguing that '[t]hey do not wish to be assimilated by the majority and they modify their behaviour to suit local conventions only in so far as this helps them to pursue their major objective', which is to save money and return home.[50] Asian families/communities are thus represented in positive terms to the extent that they are perceived as stable, but there is a negative emphasis on self-containment that develops into a construction of Asian women as seldom leaving the home and, crucially, as speaking little English. As Hazel Carby points out, these constructions were elaborated into ideologies that justified 'increased state intervention into school and home', with social workers urged to contact Asian mothers whose language deficiencies were thought to be harming their children's development.[51]

By contrast, West Indian families were seen as unstable and dysfunctional, and in an ironic twist, these characteristics were traced back to the damaging impact of slavery. Drawing on earlier studies carried out in the West Indies, a number of race relations theorists argue that slavery was responsible for the virtually complete breakdown of family structures in the Caribbean, so that legal marriage became the preserve of the upper-classes, with common law unions and promiscuity as the norm. In turn, they contend, these relationship patterns have been replicated in immigrant communities in the UK. Sheila Patterson, for example, writes of the kinship patterns of West Indian immigrants in Brixton that:

> These types of union and the family patterns to which they give rise, or which precede them, must be understood in relation to the historical and socio-economic background in the West Indies. All types of grouping but that based on legalized marriage arose primarily as a consequence of slavery, with its disregard of conjugal and kinship ties and *its encouragement of sexual promiscuity and rapid breeding.*[52]

Ken Pryce similarly argues that slavery was responsible for 'the total destruction of conventional family life', leading to the development of predominantly matriarchal family units in the West Indies and later in the UK.[53] A large number of West Indian families in post-war Britain were indeed headed by women, but the reasons for this are complex. They are bound up with the complicated intersections among race, class and gender that led to sharply differing patterns of employment among men and women. Women found it relatively easy to get employment, notably in the National Health Service but also in a wide range of domestic work, and in consequence often became the primary breadwinners for their families. The connection between the female-headed households that emerged in these circumstances and the institution of slavery is tenuous to say the least, but in making the connection, commentators such as Patterson re-animated old stereotypes about black women's high fertility and inclination towards 'rapid breeding'. The fact that many families were headed by single mothers also aroused concern because of the supposed ill-effects on their children. It was felt that these mothers could not always provide a secure environment as they often worked un-social hours (if they were in nursing, for example). Even if they did work standard hours, this prevented them from attending school meetings, which were seen as crucial points of intersection between the home and the wider community.[54] In consequence, their children were disproportionately categorised as educationally backward or psychologically maladjusted.[55]

Concerns about the perceived instability and size of West Indian families prompted the Family Planning Association (FPA) to take action. The FPA, which was first founded in 1930 as the National Birth Control Association, had close links with the Eugenics Society. Their chairman, Lord Horder, was a longstanding member of the Eugenics Society and was also the physician to the Royal Family, which says a good deal about the respectability of both eugenics and the FPA at this time. In many respects, the FPA acted as an executive arm of the eugenics movement in the post-war years, the initiative they set up in relation to West Indian families being exemplary in this respect. Concerned about the disruptive effects of large West Indian families living in overcrowded accommodation, they decided to distribute leaflets targeted at this particular population group. The campaign was run through their branches and was tailored to local circumstances, as for example in this leaflet aimed at new immigrants in South London:

> When you arrive in Great Britain, everything will be new and strange, and you will certainly want a little time to settle down before you think of starting a family or adding to the one you have.
>
> The Family Planning Association, however, with headquarters in London, wants to help you to have your babies when you want them, so they will be born when you are well and have a home and are able to take care of them.
>
> Babies are best born about two years apart, so until you are ready for the next one, go to one of our clinics, where the woman doctor will show you what to do.
>
> The address of your nearest clinic can be obtained by writing to the Family Planning Association, 64, Sloane St, London, S.W.1 or telephoning SLOane 0451.[56]

A list of local clinics followed. It is hard to know how effective such a campaign could have been, given the difficulties of getting to the clinics for women who already had children or were working. However, the FPA persisted, liaising, for example, with a vicar in South London who offered to hand out advice sheets to the West Indian couples he married.[57] There was some anxious debate over the morality of giving advice to West Indian women who were unmarried but in what the secretary of the FPA called 'free unions'. However, concerns over large families prevailed, and it was agreed that contraceptive advice should be offered regardless of relationship status.

In addition to the West Indian initiative, the FPA was heavily involved in the prescription and use of Depo-Provera, a controversial injectable contraceptive for which they produced glossy promotional leaflets. One injection of this drug prevented pregnancy for at least three months, but there were significant side effects that could include heavy menstrual bleeding, significant weight gain, nausea, dizziness and depression. Even more serious were the long term effects, which included loss of fertility and an increased risk of hormone-related cancers, diabetes and osteoporosis. Depo-Provera was licensed for use in the UK in 1974 in very limited circumstances, to prevent conception in women whose partners had just had a vasectomy or those who had just been immunised against rubella. It soon became clear that these guidelines were being flouted as the drug was being prescribed much more widely, especially for immigrants. This raised suspicions that doctors were being guided by their own prejudices in their use of the drug, and there were also concerns about the extent to which they were explaining the implications of taking it: many doubted that women were giving properly informed consent. Jean Robinson, the ex-chair of the Patients' Association, was so concerned that she became involved in a London Weekend TV programme on the issue that was shown in 1979.[58] The Brixton Black Women's Group also campaigned against the drug, as documented in Bryan, Dadzie and Scafe's *The Heart of the Race: Black Women's Lives in Britain* (1985). The issue came to a head when the manufacturers appealed against the rejection of their application for a license for long term use of the drug. A public hearing was announced, and a Co-ordinating Group on Depo-Provera was set up that brought together the various groups campaigning on questions of 'abuse, informed consent, and racism'.[59] They compiled a 48-page report for the hearing, which drew on expert medical advice from a toxicologist, a highly qualified GP and Valerie Beral, the epidemiologist who later became the director of Cancer Research UK and the 'One Million Study' on the long term effects of Hormone Replacement Therapy (HRT). The report was also endorsed by Peter Huntingford, professor of Obstetrics and Gynaecology at the London Hospital and St Bartholomew's Hospital, which served the immigrant communities of the East End of London.

The report is incisive in its analysis of the racial politics surrounding the use of Depo-Provera in both Britain and the developing world. In relation to the latter, it explores the intersections between the political interests of Western governments and the profit motives of large pharmaceuticals such as Upjohn, the manufacturers of the drug. In the UK context, it amasses evidence to show that the drug was offered to, in the words of the report, 'women who are very poor and who, overwhelmingly, are non-English speaking and/or black'.[60] The drug was only prescribed for white women if they were seen as part of the 'social problem group'. The report quotes from an article by Dr Elizabeth Wilson in which she describes giving Depo-Provera to women whose lives were caught up with problems such as 'unemployment, debt, imprisonment, truancy, homelessness, alcoholism, wife and child abuse, and degrading poverty'; Wilson also claims that most of the women were 'below average intelligence' (23). In another case, an 'openly racist doctor' injected Depo-Provera into a 14-year-old West Indian girl while she had still not fully recovered from an abortion. When the doctor was later challenged over this, he claimed that she was educationally sub-normal, despite the fact that this was not in her case notes and there was no previous suggestion of this. In yet another case, a doctor in Haringey, London, spoke of prescribing Depo-Provera for young black women who already had several children without explaining the risks, defending this on the grounds that 'the amount you tell a woman has to be graded with what you feel her level of understanding might be, I mean you can worry people very much by going into all the list of side-effects and that puts them off completely and they can't evaluate the risks' (35). The report also cites a study of two FPA clinics in the East End of London which showed that Asian women comprised two-thirds of the women on Depo-Provera in one clinic and one-third in the other. Many spoke little or no English, and the author of the study 'doubted whether more than one out of five patients had been given proper information about the effect of Depo-Provera' (32). The report concludes that Depo-Provera was being used not for birth control, but for population control. This crucial distinction is glossed in these terms: 'Birth control is about people deciding to exercise control over their own fertility. Population control means that fertility is controlled by the decision of outsiders, national governments, religious bodies, international agencies' (39).

In making this point about women's agency, the report echoes the debate around 'A Woman's Right to Choose' in the 1970s, when black women argued for the right not only to have access to free contraception, but also for 'the right to choose to *have* our children, planned or unplanned'.[61] The racial politics surrounding the use of contraception are highlighted in Emecheta's *In the Ditch*, when Adah is lectured by a Social Security clerk on the subject of large families, which are, according to him, 'not trendy' and uncivilised. Adah's response is trenchant:

> Would you, sir, consider the Kennedys uncivilized or un-trendy? What do you think of the Royal Family? Who is going to replace all those thousands of children being killed in Africa by some woolly-headed political idealists? Tell me! You dare sit there and preach to me about the world's population explosion in a place like England.[62]

Emecheta draws a number of issues together here. Adah's reference to the Nigerian-Biafran War situates her current position 'in the ditch' in its broader geo-political context. She is a Nigerian whose impoverished childhood was a direct result of British colonial policies, which were also implicated in the instability that followed independence and subsequently led to the war. In juxtaposing the killing of African children with the new concept of the 'population explosion', Emecheta underscores the parallels between the violence of war and the violence of 'population control' targeted at specific racial groups in England and elsewhere. Further, by citing the British Royal Family and the Kennedys as well-known examples of large families (or clans), Emecheta exposes the hypocrisy of stigmatising the large families of the poor (of whatever race), while those of the powerful are celebrated.

In *Second class Citizen* (1974), the prequel to *In the Ditch*, Emecheta returns to the complex problems that black women faced in attempting to take control of their fertility. In this novel, the younger Adah tries to get access to contraception because of acute economic pressures (her husband is unemployed and the family is living in one room). When she arrives at the Family Planning clinic, she sits down next to a young West Indian woman in order to be, as she says, on 'home ground'. Although she has previously expressed exasperation over the fact that the British lump all black people together ('As long as you're black, any other black person is "your people"'), in this situation, she is glad of support against the common enemy, the staff of the clinic. They are not unkind to Adah, but they treat her roughly, scolding her because she is unable to relax while they are trying to fit a contraceptive cap. And as Emecheta stresses, although the doctor and the midwife are rummaging around in the most intimate spaces of Adah's body, they have no understanding of her subjective experience of this cross-cultural encounter, in which the remembered voices of her husband and her mother-in-law cut across the advice of the nurses:

> The fear of what [her husband] Francis would say and what he would write to his mother and her relations loomed, full of doom, in her subconscious. Only she could feel it. The other two females, who were now tut-tutting at her and growing impatient and telling her to relax her legs, could not see the same picture that Adah was seeing. It was the picture of her mother-in-law when she heard that Adah went behind her husband's back to equip herself with something that would allow her to sleep around and not have any more children.[63]

Adah is caught between the traditional beliefs of her husband and his family and the values of modernity and 'population control' (in what seems like a deliberate allusion to the Depo-Provera report, which she is likely to have read, Emecheta has 'Population Control' as the title of this chapter). Both are coercive: the force of population control is signalled through the effects of the ill-fitting cap that hobbles Adah, the pain forcing her to limp around as she puts her children to bed, while that of tradition is demonstrated when Francis finds out about the cap and beats her until she is bleeding and 'dizzy with pain' (154). Adah's body, like that of thousands of West Indian, Asian and African women in Britain at this time, has become a site of biopolitical and racial struggle.

Before the Second World War, the question of race was of only marginal interest to British eugenicists, but post-war immigration changed all that, as the distant 'others' of the colonies became transformed into what Powell called the 'enemies within' and the discourses of eugenics and race were explicitly intertwined in the UK.[64] The formulation of what could be called a eugenic racism in this period was profoundly damaging in social and psychological terms, and it also impacted directly on black British women's health and rights of self-determination. Moreover, the focus on the need for 'family planning' in certain racial groups set the scene for a broader concern about birth rates in these populations. The next chapter explores the extension of eugenic racism onto the world stage, through the involvement of successive British governments with 'population control' at an international level.

5 Population Control

> Those who do not have the individual initiative and intelligence to plan and control the size of their families should be assisted, guided and directed in every way to eliminate the undesirable offspring, who usually contribute nothing to our civilisation but use up the energies and resources of the world.[1]

> Some people imagine that almost all the ills of India are due to overpopulation and therefore the basic remedy for those ills is to try to limit the growth of this population. This approach, to some extent, diverts attention from important social problems.[2]

> The things that make family planning acceptable are the very things that make it ineffective for population control. By stressing the right of parents to have the number of children they want, it evades the basic question of population policy, which is how to give societies the number of children they need.[3]

In 1952, C. P. Blacker left his post as secretary of the Eugenics Society and became the first administrative chairman of the International Planned Parenthood Federation (IPPF). The IPPF has now been reconfigured as, in the words of its website, 'a global service provider and leading advocate of sexual and reproductive health and rights for all'.[4] However, it was originally set up by a group of leading eugenicists in a period when a new sphere of activity was opening up for the eugenics movement, as both governmental and non-governmental organisations began to focus their attention on global population trends. As early as the 1930s, an informal network of eugenicists, demographers and birth control campaigners had raised alarm not only about the prospect of a declining birth-rate in Britain but also about the threat of higher population growth among non-white peoples.[5] This dual threat, of internal decline and external growth, was duly inscribed in the report of the Royal Commission on Population, which was commissioned by Winston Churchill in 1944 and published in 1949. According to the experts who submitted evidence to the Commission, the underlying trend of the British birth-rate was beneath replacement levels, whereas there was an accelerating birth-rate among 'Asiatics', which represented a threat to 'Western values, ideas and culture'.[6]

The report's recommendations were focused primarily on ways of increasing the British birth-rate through such measures as increases in family allowances, tax benefits in favour of large families and the provision of childcare facilities. The possibility of increasing the population through immigration was raised, but it was noted that newcomers would be welcome only if they were 'of good human stock and were not prevented by their religion or race from intermarrying with the host population'—in other words, if they were white. The issue raised in the report of the 'Asiatics' crowding the borders of the empire was briefly discussed in Parliament, but the government was sensitive to the political dangers of endorsing racialised concerns about the balance of the world's population.[7] These concerns were, however, taken up by activists operating through a closely linked group of non-governmental organisations based in the UK and the US, of which the IPPF was the most influential. The driving force behind the setting up of the IPPF was Margaret Sanger, the controversial birth control campaigner and eugenicist who favoured compulsory sterilisation for the 'retarded' and had in the past endorsed the idea of a hierarchy among races. However, Sanger's activities were seriously restricted by ill-health at this time, so that her role in the organisation remained largely presidential. The IPPF's head office was located in London, in offices provided by the Eugenics Society, with a small group led by Blacker making the key policy decisions. The organisation was initially funded by the Brush Foundation for Race Betterment (Cleveland, Ohio) in conjunction with the Eugenics Society.[8]

The IPPF expanded rapidly, and by 1961 it consisted of 32 affiliate organisations from countries divided into four main regions. Its focus was what was becoming known as the 'population explosion', a concept that was most often framed in terms of the transition theory developed by the influential American demographers Frank Notestein and Kingsley Davis. According to this model, excessive population was the result of the transition that occurred when traditional societies became industrialised and were able to lower their death-rate, but without a corresponding fall in the birth-rate. Transition theory predicted that the birth-rate would eventually fall, as the modernising process brought social and economic changes that would affect attitudes to fertility. However, according to Notestein, the rapidity of growth in the 'under-developed' nations had created a crisis so grave that it threatened international stability, and immediate intervention was required in the form of birth control programmes funded by the West. As Susan Greenhalgh has pointed out, the assumption that 'reproductive Westernization is good for everyone' was built into such a framing of the population problem; moreover, in constructing an opposition between patterns of reproduction in 'developed' and 'under-developed' nations, the transition model tended to reinforce the view that there were inherent differences between different population groups.[9]

However, by melding together the concepts of modernisation and birth control, Notestein and Davis provided an interpretive framework that was flexible enough to take in a wide range of interests and agendas within the population movement. This framework was especially helpful to Blacker, as he took his 'crypto-eugenic' strategy into the field of population control. Blacker defined crypto-eugenics as the 'policy of pursuing eugenic ends by less obvious means', and in this case the eugenic ends he had in mind were grounded in his belief, discussed in the previous chapter, that there were innate differences between races, and that the changing balance between races could pose a threat to the development of the species.[10] However, he never expressed this view publicly; even in his private comments on the changes caused by post-war emigration and immigration, Blacker took care to link the 'inferior' quality of immigrants to the fact that coloured people were accustomed to lower standards of living. This emphasis on the importance of the environment made it possible for him to argue for population control in the 'under-developed' nations on the grounds that if family size were restricted, the children who *were* born would benefit from higher standards of nutrition and care. He made this point repeatedly, linking birth control not only to the health of children but also to maternal health. In a 1962 article for the *Eugenics Review*, for example, he argues that voluntary sterilisation (of which he was a keen supporter) serves the health of existing children ('paediatric considerations') and the welfare of mothers. He locates this claim in the context of an alleged shift in attitudes in which birth control has been incorporated into a 'humanitarian' agenda. He writes:

> There has also been a move from a eugenic towards a humanitarian and paediatric standpoint. Sterilization, it is now widely believed, can help families. It can help mothers whose health is suffering from the combined effects of malnutrition and of numerous unplanned pregnancies—witness the numbers of puerperal sterilizations; it can relieve the father of the cares depicted in the films produced in under-developed countries which contrast the home conditions resulting from planned and unplanned parenthood; and it favours the health and life-prospects of the first two or three children. The well-being of children already born or likely soon to be born (paediatric considerations) supplement the welfare of a remote posterity (eugenic considerations).[11]

Blacker here contrasts the sufferings of large families in the 'under-developed countries' with the improved life prospects of those that embrace contraceptive modernity; only then does he make the point that such reductions in family size will improve the quality of the world population. As this suggests, Blacker was a skillful propagandist both for the population movement in general and the IPPF in particular. In relation to the latter, he was also adept in his handling of the competing demands of, on the one hand, old-school eugenicists such as Sanger and Dorothy Brush and, on the other, Elise Ottesen-Jensen, the veteran Swedish campaigner for progressive sex education. Under his direction, the IPPF central office became a 'gigantic switchboard' for activists in different countries: indeed, as Matthew Connelly suggests, the IPPF's principal asset was 'its capacity to facilitate international networking, especially between activists in rich and poor countries'.[12] The success of the operation is indicated by the fact that by 1958 Vera (later Lady) Houghton, the secretary of the IPPF, had a staff of six working in the London office.

However, Blacker resigned from the IPPF in 1960, partly in response to the increasing influence on the organisation by US representatives. In May 1960, the organisation set up a Management and Planning Committee in which, as Connelly puts it, 'the Americans secured a commanding voice in the key decisions', deciding, for example, on the allocation of grants so that they could buy influence among the affiliate organisations that received this money. This placed the IPPF in an awkward position. As Blacker argued, in order to be effective, it had to stand for universal principles such as 'family welfare, community well-being, and *international good will*', but it could only do so if it could be seen to be funded by a range of organisations, preferably also operating at an international level.[13] An obvious way forward was to seek support from the United Nations, but this proved a protracted business. Julian Huxley attempted to secure UN engagement with the population problem in 1948, when he warned the secretary-general that the world was facing a crisis that could have disastrous long term consequences. He lobbied for an international conference to consider the problem, but the suggestion was politely rebuffed. Three years later, the International Union for the Scientific Investigation of Population Problems (IUSSP) gained UN approval for a world conference, which went ahead thanks to the support of Alva Myrdal, the director of social sciences at UNESCO and (as discussed in Chapter 1) a long-time supporter of eugenics. However, this attracted little public interest, and no further conferences on this theme were held under UN auspices until 1974. Throughout this period, the UN Population Division helped to design censuses and construct projections for policy makers and activists, but it was not until 1969 that the UN created a dedicated Fund for Population Activities (UNFPA). However, this did not solve the problem of the disproportionate influence of the US on the population movement, rather the reverse. As Connelly has shown, most of the funding for population control programmes ultimately came from the US taxpayer: the cheques might be signed by the UNFPA or the IPPF, but they were underwritten by the US government.

Sir Colville Deverell, who succeeded Blacker at the IPPF, thus faced the same problem as his predecessor over undue influence from the US. Deverell was a former colonial administrator, and under his leadership the IPPF's London headquarters was perceived by some as clinging to the outdated values of the British Colonial Office.[14] Nonetheless, like Blacker, he was successful in holding the unwieldy organisation together and in reconciling the competing demands of its affiliate national organisations. Hence his resistance when in 1968 he was approached by the director of the US Agency for International Development (USAID), which had just secured millions of dollars from the US Congress, and urged to apply for an extremely large grant from them (USAID also contacted the Family Planning Associations of India and Pakistan to make the same suggestion, bypassing the IPPF head office). Deverell was outraged by this request, which in his view went against the fundamental principles of the IPPF; like Blacker, he believed that the success of the organisation depended on the independence of its national affiliates. He eventually succumbed to pressure from the US, and took the money, but he resigned shortly after. The episode throws into sharp relief the power relations that structured the field of population control. Such massive imbalances of power existed between rich and poor nations that it was possible for agencies such as USAID, the World Bank and the UN to make overseas aid contingent on the acceptance of birth control programmes, a power that, as we shall see, was exercised most forcibly in the case of India. In turn, the organisations that implemented the programmes were able to determine the nature of the contraceptive intervention that was offered, as for example when a UN team led by Deverell pushed for the use of intra-uterine devices (IUDs) in India on the grounds that, as Deverell put it, they could be 'inserted once and then [could] be forgotten about for an unlimited period'.[15] This cavalier statement disregarded the side effects for the women concerned, which could include heavy menstrual bleeding, ectopic pregnancy and pelvic inflammatory disease.

EUTOPIAS

Throughout this period, visions of population growth and its potentially fatal consequences were presented by activists in apocalyptic terms. For example, in a leaflet dating from 1957 and circulated worldwide, the IPPF warned that humanity's most spectacular achievements 'will count for little unless we can gain some freedom from one of nature's tyrannies—*the tyranny of uncontrolled reproduction which today weakens and demoralizes the family and chokes the nation*'.[16] Here the metaphor of parasitical growth pathologises the growth of the population, as does Julian Huxley's comparison between untrammelled population growth and malignant disease in an article in the popular magazine *Scientific American*. Reflecting on the sight of crowds gathered on the banks of the Ganges, he expresses the fear that 'man will turn into the cancer of the planet': the juxtaposition of the people of India and the process of disease racialises the alleged pathology.[17] Indeed, the image of the impoverished crowds of India became a staple of population discourse, as in the opening lines of Paul Ehrlich's controversial polemic *The Population Bomb*, where he writes:

> I have understood the population explosion intellectually for a long time. I came to understand it emotionally one stinking hot night in

> Delhi a few years ago. My wife and daughter and I were returning to our hotel. . . . The streets seemed alive with people. People eating, people washing, people sleeping. People visiting, arguing, and screaming. People thrusting their hands through the taxi window, begging. People defecating and urinating . . . the dust, noise, heat, and cooking fires gave the scene a hellish aspect.[18]

In this passage, Ehrlich sets up a stark dichotomy between the Western nuclear family, anxious to return to their hotel (the emblematic site of modernity, cleanliness and privacy), and the promiscuous life of the Delhi streets, in which the distinctions between the public and the private, the psychic and the somatic, are blurred and threaten to dissolve. The specific threat of racial contamination is highlighted through the focus on the speaker's young daughter, whose bodily integrity must be protected as she is conducted through this inferno made up of the bodies of racial 'others'.

The 'Indian trope' also appears in Aldous Huxley's *Island* (first published in 1962), one of a number of post-war novels that responded to the anxious projections of the population activists and engaged obliquely with the question implicit in the term 'population control': control of whom and by whom? *Island* approaches these issues from the perspective of Will Farnaby, a journalist who is touring the Pacific Islands and acting as an undercover agent for an oil and newspaper tycoon. A latter-day Robinson Crusoe, he is shipwrecked on the island of Pala, which is remote, peaceful and forbidden to outsiders. Will exemplifies many of the ills of modern civilisation: he is jaded, cynical and traumatised by the violence he has seen in Algeria and Korea during the war. His most powerful instinct is to avoid commitment, but against his will he is drawn into a struggle between the people of Pala and a dictator who wants to seize the island's valuable oil reserves. As Huxley presents this conflict, it is not between Western modernity and natural innocence, but between primitive greed and a more sophisticated approach to the relationship between man and his environment. This approach has its origins in the friendship between the nineteenth-century raja of the island and a Scottish doctor who successfully treated him for cancer, using hypnosis to anaesthetise him during a complex operation. Their friendship, and indeed this operation, act as metaphors for the curative powers of Western science combined with Eastern philosophy: it is by bringing these two forces together that the two men have 'invented modern Pala', introducing a somewhat eccentric social system that depends, among other things, on strict population control. Huxley takes care to situate the introduction of a population policy in the context of the doctor's experience of working in Madras during a famine, supplemented by Will's recent memories of the 'homeless, hopeless thousands he ha[s] seen sleeping like corpses in the streets of Calcutta'. Such representations imply that India must be saved (by the West) before it destroys itself, even though, as Aldous Huxley pointed out in a study he carried out for

UNESCO in 1948, the large populations of countries such as India were perceived by many in the West as a form of strength, a reserve of manpower that could pose a significant military threat.[19]

Population control is established on Pala through sophisticated propaganda techniques, and in this respect, Huxley's utopia turns out to be uncannily close to the dystopia of *Brave New World.* Indeed, *Island* works as a mirror image of the earlier text, drawing out elements of its fictional world and presenting them in a positive light. To take just one example: drug use and 'solidarity' rituals are obligatory for the inhabitants of *Brave New World*, and though it is often argued that they are the object of Huxley's critique, Bernard Marx for one is envious of the 'balanced life' achieved by those able to participate in the communal rituals associated with the drug soma. Similarly, on Pala, the commitment of the population to desirable social behaviour is ensured through a combination of meditation and mescalin: minds and bodies are disciplined so that individuals 'choose' what is good for the larger whole. In connection with birth control, teenagers are taught *maithuna*, a form of tantric sexual yoga that brings a heightened sense of physical pleasure with the added benefit that it prevents conception. This is so widely practised that the population is kept at replacement levels (in contrast, it is stressed, with the rapidly rising numbers in India, Ceylon and China). However, population control on Pala does not stop at the limitation of family size. All children are screened at the age of five and given blood tests, psychological tests and what is called 'somatyping', with a view to identifying individuals who pose a threat to society. The most common of these are delinquents of the 'Peter Pan' type, who are slow to develop physiologically and can become violent and aggressive. According to a doctor who specialises in treating them, Hitler was a recent example of this type, '[h]opeless at school. Incapable either of competing or co-operating. Envying all the normally successful boys—and, because he envied, hating them. . . . Thirty or forty million deaths and heaven knows how many billions of dollars—that was the price the world had to pay for little Adolf's retarded maturation'.[20] However, on Pala, the condition is now treated biochemically, Peter Pans being converted to a state of 'sweet reasonableness' through 'early diagnosis and three pink capsules a day' (152). Those belonging to another potentially dangerous type, the 'Muscle Men', are controlled through techniques of auto-suggestion. In Huxley's utopia, then, the use of biochemical and psychological techniques to shape the body politic is wholeheartedly endorsed.

The novel also endorses a programme of 'race improvement' that has similarities to the ideas of 'germinal choice' promoted by Hermann Muller and Julian Huxley, discussed in Chapter 3. The aim of germinal choice was to achieve racial uplift through the use of the sperm of especially 'fit' donors; in *Island* we see such a scheme in action through a couple who have 'decided to enrich the family with an entirely new physique and temperament'. In this case, the 'genius sperm' is that of a very fine artist: it is assumed that his talents will be transmitted to the child through genetic inheritance. The mother, Shanta, is herself the child of artificial insemination by donor (AID), her father having been selected for his academic ability. The use of AID is justified in ethical terms, when Will is told that on Pala 'most married couples feel that it's more moral to take a shot at having a child of superior quality than to run the risk of slavishly reproducing whatever quirks and defects may happen to run in the husband's family' (188). As a result, the quality of this island race is improving, particularly in relation to IQ, which is as much of a concern for the Palaeans as for the British eugenics movement. So Shanta's husband, contextualising the decision she has made, explains to Will that:

> 'Shanta has thirty two half-brothers and twenty-nine half-sisters. And more than a third of them are exceptionally bright.'
>
> 'So you're improving the race.'
>
> 'Very definitely. Give us another century, and our average IQ will be up to a hundred and fifteen.'
>
> 'Whereas *ours*, at the present rate of progress, will be down to about eighty-five. Better medicine—more congenital deficiencies preserved and passed on. It'll make things a lot easier for future dictators.' (188)

The significance of the allusion to dictators becomes clearer in the final part of the novel when, under the influence of mescalin, Will has experiences ranging from ecstatic bliss to unimaginable despair. Among the images of horror that flicker before him is a picture of Hitler's Brown Shirts marching before the war:

> Thousands upon thousands of them, their banners fluttering, their uniforms glowing in the infernal brightness like floodlit excrement. Numberless as insects, and each of them moving with the precision of a machine, the perfect docility of a performing dog. And the faces, the faces! . . . The monstrous face of Hitler with his mouth open, yelling. And then the faces of assorted listeners. Huge idiot faces, blankly receptive. (274)

In this arresting passage, the Nazis are represented in terms that echo their own anti-Semitic propaganda, for example, in the deployment of entomological and excremental metaphors. Huxley's contention is that the Nazi troops can themselves be seen as dysgenic, constituting an inferior race within a race with their 'idiocy' and slavish obedience to the Führer. Huxley was not alone in making this case: the argument was advanced by Margaret Sanger when she pointed to the concentration camps as evidence of a 'widespread devaluation of human lives' and of the consequent need for policies to improve human quality.[21] C. P. Blacker also argued for a clear distinction between 'enlightened' eugenic policies and the barbaric excesses of Nazi Germany.[22]

The concept of 'enlightened' eugenics is also central in Naomi Mitchison's novel *Solution Three* (1995), which is set further in the future and depicts a society established in the wake of a nuclear disaster. The effects of this disaster have been compounded by a rise in population leading to overcrowding and massive food shortages; in response to the crisis, a world governing body has been set up with a single overriding aim—to establish stability. To this end, they have created genetically modified crops that are standardised across the globe, with built-in virus protection. In order to limit the population, they have taken an innovative route by introducing compulsory homosexuality (an option also explored in Anthony Burgess's 1962 novel *The Wanting Seed*). The shift from heterosexuality has been achieved through a combination of biochemical and psychological treatment, reminiscent of the techniques used on Huxley's Pala. In addition to reducing the population, however, the World Council has also invested heavily in 'improving' it, on the grounds that a change in the nature of human beings is needed if the survival of the species is to be secured. In this respect, Mitchison's fiction speaks to and dramatises the hidden agenda of the population movement, which was not merely to limit the quantity of the global population but to improve its quality (although it must be emphasised that 'quality' was defined in widely different ways by the various interest groups involved).

In *Solution Three*, cloning has been adopted as a means to population uplift. The clones are derived from just two individuals, known simply as 'He' and 'She', an African-American man who played an important role in ending the 'race wars' of the twentieth century and a white woman from the Shetlands who made a 'great contribution' to the population problem while working as a doctor in India.[23] These two have been selected for their exemplary qualities, which the governing body assumes will be replicated in the clones, creating a 'superior' population. The prescience of Mitchison's novel lies in its insistence that this particular vision of cloning is an over-simplification, based on genetic determinism. The somewhat barbed dedication of the novel is telling in this respect: Mitchison dedicates it 'To Jim Watson, *who first suggested this horrid idea*' (my emphasis). Watson and Crick were genetic determinists *par excellence*, arguing that DNA was the all-powerful agent that controlled development in a top-down process.[24] As we have seen, recent work in epigenetics has demonstrated that this is not in fact the case, and that, on the contrary, as Evelyn Fox Keller writes, DNA 'is always and necessarily embedded in an immensely complex and entangled system of interacting resources'.[25] Fox Keller's perspective was anticipated in the work of Mitchison's friend C. H. Waddington, and in *Solution Three*, Mitchison draws on his theories to bring into view what he called the 'epigenetic landscape', that is, the non-DNA factors that interact with genetic material to shape human development.[26] Waddington's theories about the interplay between genes and their cellular environment are dramatised through the experience of the novel's surrogate mothers, who experience a strong sense of biological connection to their children despite the fact that their DNA was knocked out when the clones were implanted. Although this sense of connection was initially disregarded, it transpires that the connection derives from 'non-chromosomal maternal influence from the cell material and by the interchange of fluids between the foetus and its host' (99).

Mitchison utilises this strand of the novel to complicate the reader's understanding of the process of development and to highlight the difficulties attendant on attempts to intervene in this complicated process. However, she does not offer a critique of such attempts on grounds other than scientific ones: there is no sense of any ethical objection to such practices. On the contrary, the implication is that greater scientific understanding can and should be utilised in order to develop more effective eugenic interventions. So towards the end of the novel, Mutumba, the chair of the Council, is reflecting on mistakes that have been made in the genetic manipulation of plants, which prompts her to wonder if the human cloning programme ('Solution Three') might also turn out to be a failure as it has been premised on inadequate scientific knowledge. As soon as she admits these doubts about 'Solution Three', however, her mind moves on to the idea of a 'Solution Four', which would allow for more effective interventions in the human gene pool, precisely because it would be grounded in deeper understanding of the developmental process. Mitchison is not averse to the idea of such 'solutions', and her working notes for the novel are revealing in this respect, as she summarises the central question of the book in these terms: 'How can a good society work in an over-crowded world (which means, central planning)? How to annex the crisis of identity?'[27] The phrase 'central planning' is chilling, recalling the pre- and post-war regimes of Soviet Russia and China, and the idea that a 'good' society might emerge from such a formation is disquieting, as is the notion of manipulating 'the crisis of identity' in order to bind the population together. This returns us to the key issue of coercion/control. Mitchison's novel is progressive in some respects. Importantly, it is not racist, as by introducing an African-American man as one of her exemplary figures she insists, albeit rather clumsily, on racial equality in biological terms. Nonetheless, the guiding assumption of the novel is that social, political and economic problems can be addressed most effectively not by altering the conditions of human life but by altering/modifying life itself. In this respect, her work remains deeply embedded in the eugenic problematic and the belief that an understanding of biology holds the key to social advancement: her fiction displays little interest in the social and political structures that might provide an alternative route to this goal.

INDIA: THE EXEMPLARY NATION

The Indian eugenics movement developed in parallel with that in Europe, and from the 1880s onwards, there were close connections between those with eugenic interests in India and their counterparts in Britain and America. An exemplary figure in this respect is the maharaja of Mysore, who in 1920 made a handsome donation to the British Eugenics Society and corresponded at length with the British eugenicist Ethyl How-Martyn on questions of class and heredity. After being reassured by her that large families were not a prerequisite for the production of 'people of genius', he opened two government-sponsored birth control clinics in Mysore in 1930.[28] However, despite such connections and shared concerns, the Indian eugenics movement had specific interests and motivations. As Mark Singleton has shown, Indian eugenics societies began to spring up in the 1920s in response to widespread concerns about 'national degeneration—physical, moral, and spiritual', concerns that can be traced back to longstanding colonial stereotypes of Indian effeteness.[29] Eugenic ambitions became closely entwined with the growing nationalist movement, and Social Darwinist ideas were incorporated into the Aryan supremacy narratives of writers such as John Woodruffe, author of *The Seed of Race: An Essay on Indian Education* (1919).[30] As Sarah Hodges has argued, the Indian eugenics movement was also able to 'indigenise' eugenics by 'using the ancient Hindu literary tradition to claim that India's cultural heritage was inherently eugenic'.[31] Specifically, it was argued that the Hindu caste system had positive eugenic effects, a claim that made its way into the writings of many Western eugenicists, including Cyril Darlington. Conversely, it was argued that the caste system acted as a spur to negative eugenics, as the social exclusion of 'defective types' made it easier for them to be 'eradicated' (Figure 5.1).[32]

Like eugenic movements elsewhere, the Indian eugenics movement also attracted supporters with more progressive social agendas, particularly feminists with interests in maternal health and women's reproductive rights. A key figure in this respect was Dhanvanthi Rama Rau, the president of the Family Planning Association of India (FPAI). Rau was a founder member of the Women's Indian Association in 1917, a member of the All India Women's Conference and a long-time campaigner for contraception; she was also closely associated with the Indian Eugenics Society, which was founded in 1916 at Madras Presidency College, where she was then teaching. Her interests in birth control and eugenics continued to be intertwined, her perspective on eugenics being 'indigenised' through her brother-in-law Shiva Rau's association with Annie Besant, the theosophist and proponent of yoga who endorsed the view that 'a higher form of civilization' would emerge through the rapid evolution of the Aryan race. In many respects, Rau could be seen as an Indian equivalent of Margaret Sanger, and she was the obvious person to approach when Sanger decided the time had come to take the international population movement to India. India's population was growing exponentially, as was that of China, but unlike China, it was an accessible democracy with strong ties with the West. India was thus a crucial test case for the international

Figure 5.1 Dhanvanthi Rama Rau, president of the Family Planning Association of India, outside one of their clinics, 1 January 1954.

population movement; if population control could not be implemented in India, it was unlikely to succeed elsewhere.

When Rau hosted the second conference of the International Committee on Planned Parenthood in Bombay in 1952, it attracted widespread publicity and numerous messages of support. It was at this conference that the IPPF was formally established, its constitution being drafted by Blacker at Sanger's request. As Connelly points out, the conference was rapidly followed by Nehru's presentation of the world's first policy for population control to the Indian parliament. The incorporation of this policy into India's first five year plan was a significant boost for the population movement, a political breakthrough that seemed to set the seal on the success of the conference. Nonetheless, the published proceedings of the conference give a revealing insight into potential splits and divisions between the various participants. In particular, it is clear that Nehru's perspective on population control differed from that of many other delegates. In his opening message, he questioned what he saw as an over-emphasis on birth control as the solution to 'the ills of India', insisting that India's problems were due to social and economic pressures as much as over-population. He also argued, just as Marx had done in critiquing Malthusian theory, that '[i]n regard to food and other necessaries of life, there is no limit to production, if properly organised'.[33] Sanger, on the other hand, argued that the 'single principle' that could solve the 'enfeeblement and despair' caused by over-population was birth control; she also emphasised the importance of eliminating 'undesirable offspring, who usually contribute nothing to our civilization, but use up the energy and resources of the world'.[34] Her eugenic perspective was shared by many Indian delegates, including Kanialal Mareklal Munshi, the governor of Uttar Pradesh, then the most populous state in India. He argued that it was 'essential that decrepit, diseased, infirm and incurable adults should be prevented, by enforced surgical treatment, from adding an unhealthy and infirm element in our national composition': this was a duty for the state and was 'in the highest national interests'.[35]

Munshi's advocacy of 'enforced surgical treatment' foreshadows the notorious sterilisation campaign of 1975–6, which was sparked off by the instability of Indira Ghandi's government but which was in many respects the logical endpoint of Indian population policy of the 1960s. In this decade, under the influence of the international population movement, India moved towards an increasing emphasis on sterilisation as the preferred form of birth control. C. P. Blacker played a key role in this development. He was now working as chairman of the Simon Population Trust while maintaining close links with the IPPF and the Eugenics Society. He had a particular interest in vasectomy, and when his *Voluntary Sterilization: Transitions throughout the World* was published in 1963, he was hailed as an expert on the subject by, among others, Hugh Moore, an American industrialist with eugenic interests who had set up a Human Betterment Association for Voluntary Sterilization. With Moore's support, Blacker set up an international conference on the subject in New York in 1964, sponsored by another lobbying group with which Moore had connections. Both Rama Rau and B. L. Raina, the director of family planning in India, were at the meeting, where the question of 'inducements' for sterilisation was discussed. The outcome was an International Association for Voluntary Sterilization chaired by India's minister of health and family planning, Blacker having insisted that India should take the lead on the issue. Blacker also encouraged the introduction of mobile vasectomy camps in India, based on the principle that it was more effective to take birth control to the people than wait for them to seek it out. The camps were often set up in rural areas and targeted not just the poor, but 'undesirable' population groups. For example, mobile units were used to sterilise people who had been institutionalised for tuberculosis, leprosy or mental illness, which were thought to be hereditary conditions. The Dalits (previously known as the 'Untouchables') were also disproportionately targeted. In Uttar Pradesh, for example, where they made up only 29% of the population, they accounted for 41% of vasectomies and constituted an even larger proportion of those brought to the clinics by government officials. Concerns over the fertility of the poor were compounded by underlying fears about the differential fertility of Muslims. These were seldom expressed in public, but in 1967, the chief minister of Maharashtra wrote privately to Indira Gandhi arguing that 'polygamous men' should be candidates for compulsory sterilisation, on the grounds that 'any group, whose personal law allows more than one wife, may overwhelm the rest of our population'.[36]

By this time, the distinction between compulsory and voluntary sterilisation had become blurred. Given that most of the 'acceptors' of sterilisation, as they were called, were living in abject poverty, and given that the cash reward was important to them, the decision to 'accept' can hardly be seen as a free choice. The use of IUDs, which were heavily promoted in the late 1960s, was similarly problematic in this respect. As the Indian government was setting ever more ambitious contraceptive targets, the ministry of health came up with a plan to give IUDs to at least 50% of women immediately after childbirth. This move was supported by Deverell, then in charge of a United Nation's Advisory Mission to India, despite the fact that the scheme was obviously flawed in relation to the issue of informed consent, as vulnerable post-partum women were being treated by doctors who received cash bonuses if they exceeded their contraception targets. This move towards coercion at the micro level can be correlated directly with the pressure being exerted at the global, macro level, as US President Lyndon B. Johnson pressed Indira Gandhi to commit to ever more radical population policies in exchange for essential US aid.

In June 1975, Indira Gandhi was found guilty of violating election law and invoked emergency powers. During the Emergency, India became the site of a brutal population control experiment, as individual rights were subordinated to the 'right to life' of the nation. Gandhi legitimated her drastic assault on the population in precisely these terms, arguing that '[s]ome personal rights have to be kept in abeyance for the human rights of the nation, the right to live, the right to progress'.[37] Under the leadership of her son Sanjay, a eugenic campaign was unleashed against the urban and rural poor, outcastes and political dissidents, especially Communist supporters. Sanjay was able to build on the formidable machinery of birth control set up by the government, and he extended it by encouraging unlicensed entrepreneurs to set up in the sterilisation business. One such figure was Ruksana Sultana, an associate of his who was involved in an incident that became emblematic of the sterilisation campaign, the attack on the mainly Muslim area of Turkman Gate in old Delhi. As part of his modernising agenda, Sanjay linked birth control with slum clearance; accordingly, when he announced that he was going to demolish the ancient buildings around Turkman Gate, Sultana set up a sterilisation camp close by. The bulldozers moved in, and the displaced residents begged her for assistance, which she provided, allegedly, only on the condition that they provided three hundred people to be sterilised. Beggars were then rounded up and loaded into a van, and fierce fighting broke out between residents and the demolition squads, which was only quelled when live ammunition was fired by the security forces. This pattern was repeated throughout the country. There were numerous reports of young men being dragged to sterilisation camps, and villages were surrounded by police who ensured that every eligible male was taken off for sterilisation. Sterilisation, in effect, became part of the economy, as sterilisation quotas were bought and sold, while incentive payments went to doctors and 'motivators' (those who 'persuaded' acceptors to undergo the operation).

During the Emergency, millions had their reproductive rights forcibly taken from them, in a campaign that could be characterised as reproductive genocide. The leaders of the international population movement were well aware of what was going on, but remained largely silent on this brutal manifestation of 'population control'. Indeed, they ducked the issue, the World Bank stating that it had no formal position on compulsory sterilisation and the IPPF claiming in 1976 that 'it would be premature for the IPPF to issue an official statement at the present time'.[38] This was disingenuous, as these organisations were fully behind India's 'ambitious' programme and provided massive financial support for it. Funding came primarily from the US (from private donors and the government), with additional support from the Swedish government, Denmark and the UK. As before, the US money was distributed through the UNFPA and the IPPF, which still had its head office in London. Now led by Dr Julia Henderson, the IPPF provided more than 90% of the budget for the FPAI, and this was increased by 60% in 1976. With support from London, FPAI branches sterilised 80,000 people in that year.

MIDNIGHT'S CHILDREN

Rushdie's novel offers a direct response to the Emergency and the sterilisation programme, and it also re-capitulates and expands on themes that are pervasive in post-war writing: birth, re-birth and the population as a site of struggle within and between nations. As Damian Grant has noted, '*Midnight's Children* is unambiguously a nativity novel', opening with the announcement of Saleem Sinai's birth and ending with his incorporation into the dust of India, a death that is also a re-birth, in line with what Rushdie has called 'the Indian talent for non-stop self-regeneration'.[39] The central conceit of the novel, that the children born in the first hour of Independence represent the potential of re-birth for India, has been extensively discussed from a range of post-colonial perspectives.[40] However, I want here to draw on Clare Barker's suggestive reading of the novel through the lens of disability studies, to develop an analysis that focuses on the text's specifically biopolitical themes and investments.[41] As Barker points out, the children of midnight have exceptional qualities of mind and body: they are endowed 'with fantastic forms of embodied or cognitive difference'. Their powers range from Saleem's miraculous 'ability to look into the hearts and minds of men' to Shiva's capacity to crack men apart with his knees; from Parvati's 'genuine gifts of conjuration and sorcery' to more unsettling qualities, as in the case of 'a boy with the fully-operative gills of a freshwater mahaseer trout' and Siamese twins 'with two bodies dangling off a single head and neck'.[42] Such anomalous, freakish minds and bodies pose biopolitical questions about the re-born Indian nation: will the children be (literally) incorporated into it, or will they be excluded by the rationalising logic of modernity? As Saleem points out:

> Midnight's children can be made to represent many things, according to your point of view; they can be seen as the last throw of everything antiquated and retrogressive in our myth-ridden nation, whose defeat was entirely desirable in the context of a modernizing, twentieth-century economy; or as the true hope of freedom, which is now forever extinguished. (278)

Rushdie here sets up a dichotomy that structures the novel, between freedom, most saliently reproductive freedom, and economic modernity. The text is dotted with vignettes that capture the subjugation of difference and exceptionality to economic logic. For example, one of midnight's children is endowed with a beauty 'so intense that within moments of her birth it succeeded in blinding her mother'; her father is warned to keep away, but one fleeting glimpse impairs his vision. Yet once the girl has her face slashed by 'an old and ruthless great-aunt', the combination of her ruined beauty and disfigurement earns her 'a healthy living' from tourists (273–4). The implication is that in order to survive, she must conform to and mirror Western expectations of India and the Indian population. Her ruined beauty symbolises the view (which as we have seen was shared by Indian eugenicists) that the Indian population was in a state of degeneration and decline, while her profession as a beggar reflects the construction of the country as a beggar-nation, dependent on Western aid.

Saleem's exceptional gift is telepathy, which enables him to see into the minds of others and, more importantly, to commune with the other midnight's children. The invocation of telepathy has a particular resonance in this novel, which can be drawn out by reading it in the light of Derrida's suggestive essay on the subject. Exploring Freud's belief in telepathy, Derrida comments that from a Freudian point of view, telepathy is 'a question of admitting a foreign body into one's head, into the ego of psychoanalysis'.[43] Moreover, Derrida suggests that if telepathy existed, it would call into question the whole structure of psychoanalysis and the related concept of the self-contained subject/object of modernity. This is Rushdie's point, for Saleem experiences telepathy as an assault on subjectivity, which is both terrifying and exhilarating. By 'admitting foreign bodies into his head', he becomes both more and less than himself, inhabiting a space where self and other can neither be confused nor dissociated. The implication is that the children's ability to forge a multiple subjectivity might represent a break with the logic of modernity, gesturing towards what Saleem calls 'a future which would genuinely differ from anything the world had seen' (271). There is a suggestive analogy between Saleem's mind and his body in this respect. His body is a heavily freighted signifier of India, his geography teacher identifying the whole map of India in 'the face of thees ugly ape'; in addition, it echoes images of Indian children on overseas aid posters of the period. More pertinently, however, Saleem's body is a porous body, construed in terms of the incessant intake and output of bodily fluids. As a baby, he drains his mother's breasts of milk and exhausts the resources of a wetnurse, while conversely 'waste matter [is] evacuated copiously from the appropriate orifices'; as a child he produces inordinate quantities of nasal goo that tends to land on those around him, and his blood flows generously when he loses a finger. Through the image of Saleem's leaky body, Rushdie undermines the ideal of the discrete and bounded body, emphasising the inevitability of somatic mixing and merging, and suggesting that there might be a corporeal equivalent of the children's multiple subjectivity, whereby responsiveness to physical interchanges might reconfigure the experience of embodied subjectivity.

Saleem loses his telepathic powers on the occasion of his first 'drainage', when his parents take him to have his over-productive sinuses cleared. The normalisation of his body has the effect 'of depriving [him] of nose-given telepathy' and condemning him metaphorically to 'barrenness, infertility, dust' (423). Like all the sterilisations carried out during the Emergency, Saleem's castration can be referred, ultimately, to the economic battle that was going on between the US president and Indira Ghandi, with Johnson threatening to turn off the tap of overseas aid if Ghandi did not impose more rigid 'population control'. From the perspective of the Indian government, the sterilisation campaign thus had a powerful economic rationale: in the short term it placated the US government and brought in overseas aid, while in the longer term it was assumed that it would hasten the demographic transition described by Notestein and Davies and would help the country to develop its own economy. Rushdie highlights the way in which such economic arguments were internalised and played out among middle-class Indians through a narrative strand focusing on the friendship between Dr Narlikar, the obstetrician who runs the nursing home where Saleem is born, and Saleem's father. Narlikar is no ordinary obstetrician: he dislikes babies and is a birth control fanatic. He is also captivated by the prospect of economic gain in a modernised India and plans to make a fortune through an ambitious land reclamation scheme, using giant tetrapods to drain the sea. Hoping to persuade Saleem's father to invest in the scheme, he shows him a model tetrapod:

> Like a three-dimensional Mercedes-Benz sign, three legs standing on his palm, a fourth rearing lingam-fashion into the evening air, it transfixes my father. 'What is it?' he asks; and now Narlikar tells him: 'This is the baby that will make us richer than Hyderabad, bhai!' (183)

Through the details of this image, Rushdie flags up Narlikar's emblematic function: his trajectory from obstetrician to entrepreneur echoes the modernising logic that, Rushdie suggests, threatens the richness and fertility of India. In Narlikar's tetrapod, the part that resembles the lingam (usually interpreted as the phallus or symbol of the life force) is converted into a channel through which sea water can be drawn off and turned into waste: a symbol of fertility becomes a signifier of sterility-for-profit. Ahmed Sinai's participation in the scheme is represented in analogous terms. Infected by its economic logic, he confuses profitability and vitality so that when the scheme fails he loses both his money and his potency. His assets are frozen, and he tells his wife, '[t]he bastards have shoved my balls in an ice bucket!' (she discovers that he has indeed become impotent) (185). However, Narlikar's sterilising agenda is eventually challenged by a group of beggar women who represent 'the Indian talent for non-stop regeneration'. After the failure of his land reclamation scheme, he arranges for a symbolic tetrapod to be placed on the sea wall, as a kind of icon of the future. One evening, however, he discovers that his icon has been transformed:

> A group of beggar-women had clustered around the tetrapod and were performing the rite of puja. They had lighted oil-lamps at the base of the object; one of them had painted the OM-symbol on its upraised tip; they were chanting prayers as they gave the tetrapod a thorough and worshipful wash. Technological miracle had been transformed into Shiva-lingam. (244)

Saleem's castration at the hands of 'the Widow' (Indira Gandhi) comes when he returns to his 'true inheritance of poverty and destitution' in the slums of old Delhi. The location is critical as it enables Rushdie to stress the eugenic targeting of the poor; specifically, the scene reads like a dramatisation of Sanjay Gandhi's assault on Turkman Gate. Saleem is living in the slums where his birth was first prophesied, among the magicians and the 'invisible people', that is, the disabled or deformed. One morning, bulldozers appear, a 'civic beautification programme' is announced and tents with surgical equipment are set up with extraordinary rapidity for 'nasbandi' (sterilisation). As Saleem asks, what chance do the magicians (who are Communists) have against the troops who are sent in, their guns having been provided—ironically—by Moscow? The ghetto is emptied, and its inhabitants are hauled away for sterilisation. However, a special fate is reserved for midnight's children, who are dangerous not only because they are poor/degenerate/diseased, but because they represent unregulated potential and the possibility of multiple, unruly differences. For them, sterilisation is not enough, because 'there was a chance, just a chance, that such operations could be reversed'. So they are castrated: 'testicles were removed from sacs, and wombs vanished for ever' (613). Gone forever are 'the possibilities of flight and lycanthropy and the originally-one-thousand-and-one marvellous promises of a numinous midnight'. The numinous midnight also signifies the possibility of independence, which in Rushdie's reading is nowhere near realisation under Ghandi. Instead, India has been placed in a position of neo-colonial dependency, as a nation in which the human, reproductive rights of its poorest citizens are expropriated by the state and traded for overseas aid. The elite, by contrast, appears to have removed itself completely from the sphere of sexual reproduction. In a neat inversion of the transcendence of sexuality advocated by Mahatma Gandhi (involving meditation and spiritual growth), Rushdie suggests that Indira Gandhi has created a post-human elite that can replicate itself, as Saleem realises with horror that 'certain high-ups in that extraordinary government (and also certain unelected sons of prime ministers) had acquired the power of replicating themselves . . . a few years later, there would be gangs of Sanjays all over India! No wonder that incredible dynasty wanted to impose birth control on the rest of us' (551). The terminology invokes the idea of cloning-as-replication, which is a recurrent eugenic fantasy—although for Saleem and the implied reader it conjures up a nightmare scenario of endless duplications.

POPULATING THE DOMINIONS

After the Second World War, fears about the supposed decline of the British population were exacerbated by concerns over the relatively low populations of the UK dominions. The government accordingly developed a number of strategies for the transplantation of 'British stock' to Australia, Canada, New Zealand, South Africa and Southern Rhodesia, investing especially in emigration to Australia. The biopolitical rationale for post-war emigration is explicitly laid out in the 1949 report of the Royal Commission on Population. As we have seen, in their discussion, the commissioners express concern that the white population of the empire might soon be overtaken by the 'Asiatic', which could lead to a conflict on a larger scale than that of the Cold War, with the dominance of Western civilisation being undermined by the rapid growth of China and South Asia. This could create a struggle for survival between 'different peoples' (in other words, 'races') across the globe. The commissioners insist that the question 'is not merely one of military strength and security', but that it 'merges into more fundamental issues of the maintenance and extension of Western values and culture'. Tellingly, the biopolitical concept of vitality is invoked as they argue that 'the effective force' of this wider commonality depends 'on the vitality of its constituent parts, which in turn is affected by their trends of population'.[44] The commission, and successive post-war governments, were as much exercised by the biological vulnerability of British stock as by the military threats posed by post-war re-alignments.

In her important study *Whitewashing Britain*, Kathleen Paul interprets post-war emigration policy primarily in political terms, arguing that it was designed 'to provide a basis for the maintenance of Britain's international position' by boosting economic development in the context of post-war decline.[45] This was undoubtedly part of the rationale, but as Paul's discussion of government papers demonstrates, the political and biopolitical components of post-war emigration policy were closely intertwined. For example, the rhetoric of consanguinity pervades political debates about strengthening links with the dominions, with frequent references to kith, kin and blood. The empire is represented as a family with a shared biological heritage, an idea further developed through the metaphor of blood transfusion (which depends, of course, on the compatibility of blood groups). So in 1954, when the Conservative Commonwealth Relations Secretary Lord Salisbury was arguing for the renewal of the Empire Settlement Act, he put it to Parliament that emigration to the dominions should be understood as 'a transfusion of blood which strengthens the whole'; equally when his successor was deploring the cost of the assisted passage scheme to Australia, he nonetheless endorsed the idea that the Commonwealth needed 'a blood transfusion which can best be supplied from this densely populated island'.[46] The image of the Mother Country as a blood donor for family members extends the concept of the British body politic, which is now understood in terms of ties of blood and inheritance stretched out across the globe.

As the deployment of this metaphor suggests, the costs of emigration were widely understood to have biopolitical as well as economic implications, with politicians and political commentators, as well as prominent eugenicists, debating the issue in these terms.[47] For example, in 1949, the journalist Dudley Barker published a polemic *People for the Commonwealth: The Case for Mass Migration* with an endorsement by Hugh Dalton, the chancellor of the exchequer in Atlee's government, in which he makes the radical suggestion that the government should 'transplant' whole towns to the dominions. This would allow the dominions to develop their population from the British people who are, he contends, 'the most skilled industrial workers in the world and who have a standard of intelligence considerably higher than the average anywhere else'.[48] In addition, through moving entire towns, the threat of competition between the UK and the dominions for the most active and talented people would be resolved: rather than just creaming off 'the healthiest and the hardest working', the dominions would take 'the old and the handicapped' as well, with these non-productive citizens contributing to the new economy as consumers. Although the scheme was never implemented, it gives an indication of the terms of contemporary debates and of the ambitiousness and scope of population planning at this time. And there was a large-scale exodus from Britain between 1945 and 1970: specifically, more than one and a half million British citizens left for Australia between 1945 and 1970, most of them travelling under the assisted passage scheme in which adults paid £10 and children travelled free (Figure 5.2).

This was one of the largest movements of population of the twentieth century, and it was impelled to a considerable extent by the eugenic investments of both the British and Australian governments. As Alison Bashford has shown, by the mid century, Australia had a well-established eugenics movement, unsurprisingly in the context of the notorious 'White Australia' policy that dated back to the founding of the nation in 1901.[49] As Laksiri Jayasuriya points out, the power of the East was already causing concern at the turn of the twentieth century, hence, he argues, the new nation of

Figure 5.2 Three brothers emigrating to Brisbane under the 'Bring out a Briton' sponsorship scheme, 1 June 1957.

Australia 'was defined, foundationally, *against* "Asia"'.[50] With the bombing of Pearl Harbour in 1941, fears of a Japanese invasion revived anxieties about 'the Asiatic hordes', and the then prime minister invoked the language of eugenics to re-assure Australians that their nation would 'remain forever the home of sons of Britishers who came here in peace in order to establish . . . an outpost of the British race'; he also spoke of the horror the Japanese inspired in him because of their 'incompatibility of race and blood'.[51] After the Second World War, the Australian government continued to uphold the White Australia policy and put considerable pressure on the UK to 'export' its people, turning to other European sources only when the British supply began to dry up. As Gwenda Tavan has shown, eugenic assumptions about the superiority of the 'British' or 'white' races continued to permeate Australian culture in this period, although several church groups challenged the White Australia policy and an Immigration Reform Group was set up in 1960, publishing an influential polemic *Control or Colour Bar?: A Proposal for Change in Australia's Immigration Policy.* By this time, the government was becoming more open to change, largely because it could see the political and economic damage the White Australia policy was causing, souring relations with India, China and Indonesia. In the late 1960s, the policy was gradually dismantled, though this was done in a secretive way due to fears of a domestic backlash.

The underlying premise of the post-war emigration policy, from the perspective of both governments, was that the white British and Australian populations constituted a single 'race' that needed to be strengthened in order to meet the challenge of a growing 'Asian' population. However, the notion of the homogeneity of the populations was undercut by a concept of Australian exceptionalism, which, as Stephen Garton has shown, had its origins in the 1920s when the historian George Challis Wood published a re-assessment of the convicts transported to Australia.[52] Wood suggested that far from being the dregs of the British criminal classes, they were vigorous and enterprising labourers forced to turn to crime as a result of the changes brought about by the Industrial Revolution. For Wood, then, the white people of the Antipodes were not inferior stock but were inherently more vigorous and enterprising than other peoples. Although Wood's study was published in 1922, the idea of the vigour of the early white settlers continued to inform the cultural imaginary and played out in a particularly negative way in relation to post-war emigrants from Britain. There was a relatively high return rate among British emigrants, and as this represented a poor yield on investment for the Australian government, it funded a number of studies into its causes. The 1967 report of the Immigration Advisory Committee concluded that many British migrants did not have 'the necessary character' to become successful immigrants, a conclusion that reflected and perpetuated the popular view of the post-war emigrants as relative weaklings. As one returning emigrant wrote in 1957, '[i]t is quite common to hear . . . that the returned emigrants are not of the same calibre as the old settlers. "They can't take it!" is a commonly-used expression about them'.[53] In consequence, many returning emigrants felt a lasting sense of failure that was linked to a sense of not being of the 'best' biological type. The sense of a biological faultline between the two populations was exacerbated by the strict screening of British emigrants before they entered Australia. There were particular concerns that they might be infected by tuberculosis, a disease usually associated with the poor that had a much higher incidence in the UK than in Australia; all British emigrants had to have a chest X-ray before being allowed into the country.

ELIZABETH JOLLEY: TRICKED OR TREATED?

'Tricked or Treated' is the title of a chapter in the Australian writer Elizabeth Jolley's memoir *Central Mischief* (1992), in which she reflects on her voyage from England to Australia when she emigrated with her young family in 1959.[54] As this title suggests, her response to emigration was one of profound ambivalence, and in another essay in this volume, she writes of the secret hopes of her fellow travellers and the many subtle forms of disappointment that awaited them. Her most searching exploration of the migrant experience is in the novels published in one volume as *The Vera Wright Trilogy* in 2010: *My Father's Moon* (1989), *Cabin Fever* (1990) and *The Georges' Wife* (1993). These autobiographical fictions have two closely linked themes at their centre: the experience of the Second World War and the post-war emigration programme. For Vera, the traumatic experience of the war is one of the factors that drive her to emigrate, as was the case for many others, according to Hammerton and Thomson in their oral history *Ten Pound Poms*. One striking case they discuss is that of Albert Lougher, who returned to Britain after being a German prisoner of war and felt profoundly alienated from post-war British society. He described his situation in this way: '[t]he people that I respected were dead. My brother had died and he was two years older than me, and the people that went away with me were all dead. I mean, what the hell was I doing back there? It wasn't my world; I should have died when they died'.[55] Vera is similarly traumatised by her war-time experiences. She works as a nurse in a large military hospital, where she witnesses appalling injuries that she is powerless to treat in any effective way. The trauma of civilians was not widely acknowledged in this period, however, and it is not until Vera is on her way to Australia years later that she realises that she has 'my own share of deep wounds to which I am responding in secret'.[56]

Vera's wounds are psychological but derive from her intimate contact with somatic wounding, an experience that is most powerfully conveyed through the recurring image of amputation. The treatment of amputees is at first described in a detached manner reminiscent of rote-learning from a nursing textbook:

> In the theatre recovery ward there are fifteen amputations, seven above the knee and eight below. The beds are made in two halves so that the padded stumps can be watched. Every bed has its own bell and tourniquet. (56)

> The ward where I am is all split beds for amputations. Every bed has its own electric bell and its own tourniquet, some beds have two. (106)

However, after the second reference to the split beds, the repressed or disavowed trauma breaks into the text as Vera remembers a young man who has had both legs amputated above the knee and also has a terrible wound in his stomach: 'I try, while I am dressing it, not to show that I can't bear it. . . . As I swab the wound it seems something is moving in it. It is a maggot. I pick it out quickly with the forceps trying not to show my shock. Suddenly I see there are maggots everywhere. It's as though he is being eaten alive' (107). Such scenes of living death prompt hallucinations, as when Vera is sent out on an errand and imagines that the darkness is full of 'rotting arms and legs, thrown out of the operating theater' (61). The trope of the broken body signals the violence directed towards individual bodies and the body politic during the war and also anticipates the trauma of Vera's post-war emigration.

Vera's self-reconstruction after the war is made difficult because of her complex 'coming of age' experiences during her time in the hospital. She is involved in a complex web of relationships that include a passion for an enigmatic female staff nurse and an entanglement in a ménage à trois, which leads to her illegitimate pregnancy. The narrative is elliptical in its account of the currents of desire between the characters, but the text emphasises the dangers of straying beyond the boundaries of normative heterosexuality. For example, Dr Metcalf, the father of Vera's child, goes off in pursuit of a male lover and dies unexpectedly in circumstances never fully explained. The ménage à trois is a recurring pattern for Vera: she becomes involved with the Georges (brother and sister, Mr George being the father of her second child) and subsequently with a pair of artists (again brother and sister), whose dissolute and incestuous lifestyle is represented in the symbolic logic of the text by the tuberculosis that Vera catches from the brother. Through this repeating pattern of ex-centric affiliations, Vera thus places herself outside normative family structures, opening up new possibilities of love and relationship but at considerable cost. She establishes a sexual and reproductive identity that is considered deviant by some (including the hospital authorities), and in terms of the norms and conventions surrounding the reproduction of the nation, she recognises that she and her daughter will never be among 'the chosen ones'. This point is underscored in the following passage, as Vera walks through the suburban streets, watching the lives of the bourgeoisie from the outside with her cold and hungry daughter:

> The particular garden, to which I am always drawn and which, particularly in summer, suggests all that I wish for, has today a snowman in the middle of the trampled snow. The lights are on and the long uncurtained windows offer a glimpse of the magic of the life within; the desirable warmth and harmony of a family, and the idea of having been especially chosen to make this family. (299)

The idea that Vera should emigrate to Australia is first mooted when she is being treated for tuberculosis; it is suggested that Australia is a country that has 'possibilities' and that it will be good for her health. However, as the ship nears shore, she is subjected to the rigorous 'offshore' screening procedures whereby Australia ensured (as it still does) the quality of its prospective inhabitants. Australia is not the free and open space of her imagination, but is a country with its own biopolitical investments, which include a specific concern with eliminating tuberculosis. Vera's history as a tuberculosis patient threatens her status as a migrant, as she realises when she sees these signs posted all over the ship:

> MIGRANTS QUEUE HERE MIGRANTS CHEST X-RAY
> NO MIGRANTS PAST THIS POINT
> MIGRANTS REPORT HERE
> (489)

The process of scrutiny and the weeding out of undesirables by the Australian authorities is further underscored when the ship leaves Bombay and a woman who was once 'the most striking Madam in the red-light district of Poona' dies and is buried at midnight:

> She slides in a secret launching along the silver path the moon makes on a sea, so silent and smooth the surface seems solid as if beaten from precious metals. The Union Jack which, with her last words it is said, she insisted upon, stays afloat, rippling, long after the unwieldy parcel sinks.
>
> The way in which the shipboard corpse is handled reminds me of the log lift ambulance men use when passage for a stretcher is not possible. (501)

In this passage, the Union Jack, the signifier of citizenship on which the un-named woman has insisted as the final marker of her identity, floats free of her body and after a while is fished up and stolen by another passenger, the incident pointing to the ultimately illusory nature of such national identifications. This does not detract, however, from the exclusionary violence enacted in the name of the imagined communities of nations. The Madam is associated with moral contagion and the threat of syphilis, so that she would have been excluded had her death not intervened; similarly, an Anglo-Indian teacher and his family (known to the other passengers as 'tinted folk') are wrongly thought to be linked with a case of smallpox on board ship, which seriously compromises their chances of entry to Australia. Vera is not excluded in any direct way, but she feels that through her experience of migration, she has suffered a loss that she describes in terms of bereavement: she has been 'widowed from one style of life to another'. She has an affair with another woman during the voyage, and they call each other 'the Widow' and 'the Migrant', the generic terms signalling their sense of being subsumed into demographic categories for the purposes of the journey. For Vera, the experience of being categorised as a 'migrant' diminishes her sense of identity, as she perceives that she is not valued as an individual but as a unit in the movement of 'British stock'. At the same time she is aware that the part she plays in this movement of population is in a sense fraudulent, as she does not conform to 'healthy' norms of social and sexual behaviour. The experience of migration thus intensifies rather than alleviates the feelings of exclusion and inadequacy that prompted her to make the journey, and the final section of the trilogy conveys her lasting sense of displacement, expressed through the metaphor of homelessness:

> I suppose homelessness, the sense of not having a place of sanctuary to return to, can come at any time, even in such separate places as the Great Bitter Lake where a ship might wait for a passage through the Suez Canal, or in the quiet suburban streets where the winter trees are finely drawn on the cold rosiness of the winter sunset. (550)

CODA

This chapter has explored the ways in which the British eugenics movement extended its reach through its involvement with the global population movement, and it has focused on two 'case histories' in order to explore the development of policies aimed at regulating the racial balance within the world population. As we have seen, a number of British eugenicists played a part in the implementation of negative eugenic policies in India, and they were also involved in advising population control movements in Ceylon (Sri Lanka) and Japan. In relation to positive eugenics, the convergence of the eugenic interests of the British and Australian governments facilitated an emigration policy that involved the mass migration of 'Anglo-Saxon stock', and similar schemes were set up in relation to New Zealand, Southern Rhodesia, South Africa and Canada. The global interests of the post-war eugenics movement have been overlooked in part because they were articulated and advanced 'offshore' via unelected and unaccountable non-governmental organizations (NGOs), but as the fiction of Rushdie and Jolley attests, they had real and damaging effects. The power of global capitalism is currently creating analogous problems as the biotech industry works across national borders and for this reason can elude regulation aimed at preventing, for example, the cloning of human embryos for research aimed at genetic enhancement. Such companies do not have eugenic ambitions, but the techniques they develop may have eugenic effects: these issues among others are addressed in the afterword.

6 Afterword

This study has explored the persistence of eugenic thought in Britain between 1945 and 1979, arguing that it did not fade into obscurity in this period but in many respects gained a new lease on life. Eugenic concerns with the health of the population overlapped with the aims of post-war reconstruction to a significant extent, and, as we have seen, some of the key architects of the Welfare State had close ties with the eugenics movement. In consequence, eugenic ideas had a demonstrable influence on aspects of post-war government policy. For example, Cyril Burt's work on IQ testing was critical to the argument for a tripartite education policy, while C. P. Blacker's analysis of 'problem families' contributed to the development of policies relating to child development and institutional care. In addition, eugenic thought had an impact on a number of academic disciplines, including social medicine, sociology and, most notably, genetic science. As Blacker argued in *Eugenics: Galton and After*, eugenics was a complex subject that involved 'policy, science and sentiment', and a key strand in this book has been an examination of the sentiments, or what Raymond Williams would call the 'structures of feeling', associated with eugenics.[1] As we have seen, these can be traced not only in literary and popular fiction but in a wide range of books directed at the general reader in the fields of biological science, psychology, sociology and social policy.

The strength of the eugenics movement in the post-war period was closely bound up with the dominance of the Welfare State, for as Foucault suggests in *The Birth of Biopolitics*, guaranteeing the health and welfare of the population was a major aim of the British government in the wake of the Second World War and the Depression.[2] Support for the Welfare State and for welfare-capitalism remained strong into the 1970s, but when a Conservative government was elected in 1979 led by Margaret Thatcher, there was a decisive shift towards neo-liberal economics and a gradual rolling back of the Welfare State. Government rhetoric ceased to invoke collective ideals, promoting instead the virtues of individual effort, the entrepreneurial spirit, market forces and consumer choice. This shift in social values coincided with the direction of genetic science, which moved away from a concern with the population as a whole and focused on uncovering the genetic make-up of individuals. In consequence, as Hilary Rose has argued, clinical genetics no longer saw itself as leading a eugenic programme on behalf of the state, but developed a professional ethics focused on patient choice supported by non-directive genetic counselling.[3]

In the context of the triumph of neo-liberal economics and the pressures of globalisation, the state gradually withdrew from many of the responsibilities associated with protecting the health of the population. As Nikolas Rose writes, the nation state no longer organises its subjects into a body politic 'on which it works en masse, in relation to the body politics of other states competing in similar terms'.[4] The state is now seen as a facilitator, encouraging individuals to take responsibility for their own well-being through numerous health promotion campaigns. Choices made possible by advances in reproductive technology are also made by individuals rather than the state, and it is on these grounds that they have been defended by a number of thinkers from the charge that they could have eugenic implications. Nikolas Rose, for example, regards reproductive choices such as those that follow pre-implantation diagnosis as 'shaped by forms of self-government imposed by the obligations of choice, the desire for self-fulfillment, and the wish of parents for the best lives for their children', and argues that because they are not directed at the health of the population, they cannot be considered in any way analogous to Nazi eugenic policies (69). Similar arguments have been advanced by Nicholas Agar in his defence of liberal eugenics: he also draws a sharp distinction between liberal eugenics and the coercive policies of the Nazi regime, and for him as for Rose the question of individual choice is central to the argument. He contends that if Nazi eugenicists had access to genetic technologies they would have used them to curtail choice, whereas liberal eugenicists seek to use these technologies to significantly extend reproductive choice. His view is that it is reasonable for prospective parents to ask for embryos possessing combinations of genes 'that correspond with their particular conception of good life', as long as they acknowledge the right of their fellow citizens to make different eugenic choices.[5]

When the Human Genome Project was nearing completion in the late 1990s, a team of distinguished bioethicists was commissioned to debate the moral issues raised by advances in such genetic technology. Their conclusions are in broad agreement with those of Rose and Agar, although they are more explicit about the overlap between eugenics and the genetic interventions they consider and less concerned to draw a *cordon sanitaire* around the new eugenics in order to distinguish it from the old. In their report *From Chance to Choice* (2000), they begin by defining eugenics broadly as 'concern for human betterment through selection—that is, by taking measures to ensure that the humans who do come into existence will be capable of enjoying better lives and of contributing to the betterment of lives of others' and suggest that this is an unexceptionable aim.[6] Arguing that past Nazi abuses should not lead to a condemnation of all aspects of eugenics, they go on to consider how the pursuit of genetic health and enhancement can be reconciled with the requirements of justice. They argue that genetic intervention to prevent disease is legitimate if the disease is likely to seriously limit an individual's opportunities in life, although they are careful to emphasise that such interventions should not be forced on individuals. They also argue that access to genetic enhancements will need to be regulated to prevent the exacerbation of existing inequalities (101). They conclude by offering a cautious endorsement of a public policy of eugenics in the broadest sense, arguing that '[t]hrough its democratic processes, a liberal society could decide to devote resources to the continual enhancement of desirable human characteristics—to embark on a process of genetic perfectionism—so long as in doing so it did not compromise its commitment to justice and the prevention of serious harm' (345). In their view, then, eugenics is not objectionable *per se*, and the prospect of genetic improvement should not be dismissed but examined critically in the light of past errors.

One of the most influential critiques of contemporary liberal eugenics has come from the philosopher Jürgen Habermas, who has argued that pre-natal genetic interventions have the potential to disrupt our self-understanding and threaten the wider social contract. His argument is that systems of religious belief and secular thought have until now been grounded in the assumption that 'the genetic endowment of the newborn infant . . . lay beyond any programming and deliberate manipulation on the part of other persons'.[7] This assumption depends on what he calls 'the rather ordinary contingency' of human reproduction, in which the outcome is uncontrollable and unforeseeable. However, once adults are able to treat the genetic traits of their descendants as a product which they can shape according to their own design, they are exercising a control that intervenes in 'the somatic bases of another person's spontaneous relation-to-self'. Habermas argues that we shape ourselves as individuals by engaging critically with our social conditioning, and what concerns him is that it would not be possible for a genetically manipulated adult to engage critically with (and/or repudiate) their biological conditioning (14). His argument has been criticised on the grounds that it rests on an untenable distinction between those whose biology is 'naturally' given and those whose biology is 'designed'; a related criticism is of his genetic essentialism, that is, the assumption that altering a person's genes would alter their identity in a fundamental way.[8] While accepting these criticisms, Michael Sandel has, however, highlighted a further implication of Habermas's contention that in order to understand ourselves as free, we must be able to ascribe our origins to a beginning that eludes human control. While for Habermas this is important in relation to the question of individual autonomy, Sandel suggests that it has a broader significance, arguing that 'the drive to banish contingency and to master the mystery of birth diminishes the designing parent and corrupts parenting as a social practice governed by norms of unconditional love'.[9]

A further critique of contemporary eugenics has come from the sociologist Troy Duster. In his classic study *Backdoor to Eugenics*, Duster moves the argument away from questions of individual choice to explore what he calls, in a memorable turn of phrase, 'the subterranean eugenic aspects of the revolution in molecular biology'.[10] As he demonstrates, with the construction of genetic explanations for illness over the last two decades, new target populations have been created for genetic interventions. He focuses in particular on the way in which the concept of a genetic pre-disposition to disease has rehabilitated the idea of race as a biological category. For example, the higher incidence of sickle cell anaemia among the black population and the implementation of screening programmes for the disease have reinforced the perception that African-Americans have a distinctive genetic inheritance, despite the fact that, as Duster points out, sickle cell anaemia is also found in a range of populations around the Mediterranean. An increased risk of heart disease and cancer, also attributed to a genetic pre-disposition, has recently become another marker of the supposed 'biological' specificity of the black population of the US. Turning to issues of social class, Duster highlights the tendency to over-emphasise the genetic component in a range of conditions associated with lower socio-economic status, for example, spina bifida, a neural tube defect with both genetic and environmental components. He argues that to name such conditions as genetic has the effect of setting in train interventions aimed at preventing the problem, such as pre-natal screening, whereas if the condition were characterised as multi-factorial, the approach might be very different. This is particularly true when (as in the case of spina bifida) there is clear evidence of the effectiveness of interventions such as dietary folic acid supplementation in reducing the incidence of the condition.

Duster's anxieties for the future are concentrated on pre-natal screening and negative eugenics—the 'backdoor to eugenics' of his title. He suggests that there is likely to be a broad consensus that pre-natal screening should be offered for life-threatening genetic disorders. However, his concern is over public policy in relation to screening for disorders that are not life-threatening but make for lifetime dependency, or those that are 'simply unaesthetic' (e.g. spina bifida) (130). His fear is that there is 'a possible new attitude waiting to be adopted—that "the defective fetus" can be eliminated', and he calls for a vigorous and informed debate on the eugenic implications of genetic screening. In the UK, this debate has focused principally on pre-natal screening for Down's syndrome. A study published on this subject hit the headlines in 2009 when it showed that the high uptake of screening followed by terminations had reduced the numbers of babies born with Down's syndrome, despite the fact that the numbers of Down's pregnancies rose by 71% between 1989 and 2008 as women chose to have children later.[11] This finding has led disability activists to argue that the screening programme makes disabled people with similar conditions feel that their validity as members of a society is undermined.[12]

New reproductive technologies and their eugenic implications have proved fertile ground for writers and film makers, who have responded to the questions they raise in powerful and imaginative ways. Eva Hoffman's novel *The Secret*, for example, explores an extreme instance of the kind of reproductive choice endorsed by Agar. The main protagonist, Iris, is a clone, something she and we discover quite early on, so that the rest of the novel constitutes an innovative *bildungsroman*, as Iris seeks to distance herself from her mother/twin in order to self-critically author her own life. One of the most striking aspects of cultural representations of clones is the way in which they depart from the scientific 'truth' about this reproductive technology. Such representations assume that the clone is an exact replica of its original despite the fact that biological cloning to date, as in the case of the creation of Dolly the sheep, involves a multiplication of hereditary and environmental factors taken from three maternal 'parents'. Dolly was created from the nuclear DNA of one ewe, the mitochondrial and cellular DNA of another and was nourished by the uterine environment of a third surrogate mother. Why, then, the emphasis on replication and similitude? For Hoffman, the trope of the clone-as-replica signifies an extreme instance of the coercive reach of reproductive choice. Elizabeth's decision to clone her daughter is presented as a solipsistic self-projection into the future, as Elizabeth herself seems to acknowledge when she tells her daughter, 'I guess I had enough vanity, or self-love, to think that whoever you were, whoever you turned out to be, you might be pleased to be . . . like me'.[13] Conversely, when Iris discovers that she is a clone, she feels at one and the same time that she has no identity (because she is not singular) and that she is alien, monstrous, a self-generated freak: 'I peered into her and was sucked into her eyes, and saw something from within them, saw her desire for my birth, for another her/me. . . . Did that mean I'd given birth to myself?' (66). In this respect, Hoffman dramatises the kind of asymmetrical relationship between parent and child, or designer and designed, which Habermas sees as one of the dangers of the new reproductive technologies. The text also opens up new ground in drawing attention to the ways in which cloning and other reproductive technologies pose a challenge to genealogy in its traditional sense, that is, to our understanding of parenting, blood ties, kinship and descent. This is not to suggest that genealogy is identical to the truth of biology, for as Sarah Franklin points out, although genealogy cannot be separated from its biological aspects, these do not determine the social forms it takes.[14] Nonetheless, Hoffman foregrounds the ways in which Iris's cloning disturbs the normative structure of kinship and challenges conventional assumptions about patterns of inheritance and descent. Merely by looking at her mother, Iris is confronted with the vertiginous possibility that Elizabeth can be 'read' simultaneously as her mother, her sister and her daughter. Elizabeth, however, wants to resist these multiple readings (and their subversive implications), just as Iris's grandparents reject her suggestion that they might be considered her parents as well as her grandparents (134).

Hoffman's novel also explores the wider social context in which Elizabeth's individual decision is made. In Hoffman's imagined North America of 2025, liberal eugenics is available only to the rich—in other words, this is liberal eugenics for a neo-liberal economy. In this world, biotech companies have no therapeutic dimension but trade solely in enhancement, aiming to produce individuals who will have 'a productive, fulfilled life' (99). But as Hoffman's novel indicates, as our individual and collective identities are increasingly understood in terms of somatic status, biosocial communities may emerge that are organised around illness/disability and also around enhancement. Iris finds a number of such online communities, and in dramatising the debates on their websites, Hoffman deliberately underscores the connections between the old eugenics and the new. On 'Clone Power', on the one hand, one clone contributor writes:

> There are only a few of us so far . . . but we are faster, smarter, more beautiful. Over time, we will select from among ourselves the most efficient prototypes. The human race is senescent and ready for a cleansing and rejuvenation. (190)

For the 'echt humans', on the other hand, the clones are either 'the hidden disease. . . . Like the Jews were in Germany' or alternatively they are like the Nazis, carrying 'the virus of eliminationist superiority' (191). These debates are, as Iris points out, 'tired', but they are also dangerous, pointing to the fissures that could open up in a society stratified according to the possession of biocapital, that is, those biological qualities that are highly valued socially and hence economically. Hoffman anticipates and warns against such a society, in which there is differential access to enhancement and individuals or groups perceive themselves as being so different that they no longer recognise others as fellow citizens.

Kazuo Ishiguro's novel *Never Let Me Go* (released as a film in 2011) similarly explores the development of social groups organised around biological differences. The text offers an alternative history of Britain in the 1990s, in which human clones are mass produced for the purposes of organ harvesting. They constitute an underclass whose biocapital is expropriated for use by the 'normals' whose lives are extended by transplantation. By contrast, the lives of the clones are brutally curtailed, as they die (or 'complete' as the narrative euphemistically puts it) as soon as they have 'donated' four vital organs. The clones' bodies are created only to be dismembered, and their subjectivity follows a similar trajectory: by this means Ishiguro points to the alienation of labour in a biocapitalist economy. As valuable capital, the clones are initially encouraged to develop a 'normal' subjectivity. They grow up in a boarding school where they are encouraged to express themselves creatively and forge close relationships with each other. However, when they become adults, they are confronted with a different kind of self-fashioning, as they are required to divert their energies and ambitions away from their singular futures and invest in their professional role as donors and carers for others. They become disengaged from their emotions and move towards a detachment from self, which is expressed in the affectless register of their speech. Matthew Eatough interprets this shift in terms of a positive move towards a 'transpersonal' realm of identification with others, but it can also be viewed as an expression of psychic withdrawal and dissociation.[15] Moreover, the clones come to regard their own bodies with a detachment that borders on indifference. While Eatough sees this disidentification in positive terms, as the pre-requisite for a caring investment in the bodies of others, the severing of the link between the affective self and the body can also be seen as a form of violence, preparing the clones for passive acceptance of their ultimate disembodiment.

Ishiguro also explores the relationship between the parent and clone in ways that resonate with Habermas's insights and extend them dramatically. In a reversal of normative patterns of inheritance, Ishiguro suggests that the clones' future lies in the past, in a point of origin that simultaneously offers the prospect of a vicarious future. Because their parents will outlive them, the clones invest imaginatively in *their* future, in a grotesque reversal of the conventional relationship between parent and child. It is for this reason that the clones search obsessively for their parents, who are also known as 'possibles', a word that underscores the sense of potential associated with them. Ruth, for example, wants to be an office worker and is captivated by the sight of a 'possible' working in a modern office block:

> There was a big glass front at street-level, so anyone going by could see right into it: a large open-plan room with maybe a dozen desks arranged in irregular L-patterns. There were the potted palms, the shiny machines and swooping desk lamps. People were moving about between desks, or leaning on a partition, chatting and sharing jokes.[16]

It is the very mundanity of this scene that has such an effect on Ruth, this and the fact that what lies behind the glass is 'a smart, cosy, *self-contained* world' as far removed as possible from the dismemberment and death that await her.

Ishiguro is explicit about the eugenic potential of twenty-first-century biotechnology. The clones are created from the 'dregs' of society, from junkies, prostitutes, winos and tramps: they and their parents constitute an abjected underclass, and it is this that makes their exploitation and extermination possible (164). Ishiguro stresses the way in which their parents (like the 'social problem group' of the 1950s) are constructed as a diseased part of the body politic, as 'trash' that can be reclaimed and recycled through the 'donor programme'. The clones, too, are constructed as 'less than human' simply by virtue of being clones and are 'kept in the shadows'. People prefer to think that the organs grow in a vacuum; disavowal enables the donor programme to flourish. Paradoxically, the conspiracy of silence around them is broken when the normals who appropriate their organs are threatened by a development in positive eugenics. In a glancing allusion, perhaps, to the eugenic ambitions of James Watson, Ishiguro invokes a scientist called James Morningside who directs a secret research facility aimed at offering people 'the possibility of having children with enhanced characteristics' such as superior intelligence. The institute is shut down because the prospect of a genetic superclass (a super race) terrifies the normals; in an echo of the concerns dramatised in Wyndham's *The Chrysalids*, they see the superclass as a potent enemy within.

Whereas Ishiguro explores a reconfiguration of the British class system in which biocapital is the most valuable possession, Margaret Atwood's *Oryx and Crake* focuses on the intersection of advances in biotechnology and global capitalism. As a number of commentators have argued, biomedical research has now become inseparable from the drive to generate intellectual property, with the consequence that illness and health have become major fields for corporate activity and the generation of shareholder value. The bioeconomy is one in which, as Nikolas Rose suggests, 'vitality' is broken up into a series of objects (tissues, blood, DNA samples) that are 'accumulated, exchanged, traded across time, across space, across organs and species' (38). Atwood's novel *Oryx and Crake* models some of the consequences of these developments. In the world of this novel, biomedical scientists work on programmes directed at both the treatment of disease and the enhancement of human life. The staff are recruited for their scientific ability, which is focused exclusively on the extraction of profit from the currency of vitality. They can be seen as examples of what Rovirosa-Madrazo dubs 'the genetocracy', which is likely to emerge in the future: the compounds where they live are 'awash in brilliant genes', and any inhabitant who fails to measure up to these high standards is discreetly excluded.[17] The main protagonist, Jimmy, for example, is a 'neurotypical' and so is sent away to a lowly arts college located in 'the pleeblands'. The pleeblands share the characteristics of many disadvantaged urban areas today—dingy flats and apartments, burnt out cars, poor sanitation and the constant threat of violence. But the factor to which Atwood draws particular attention is the way in which the privileged few have been taught to see the 'pleebs' as forming a genetic underclass. This is not primarily an issue of physical fitness or its absence, although the pleeblanders have strange 'asymmetries and deformities'. It is primarily a question of that quality which has been fetishised by eugenicists from the time of Francis Galton onwards: intelligence. Jimmy assumes that 'nothing of interest went on in the pleeblands . . . there was no life of the mind', and he remembers hearing the old term 'mental deficient' applied to the pleebs.[18]

In *Oryx and Crake*, the biosocial stratification of *Never Let Me Go* operates on a global scale, as developments in biotechnology are driven by the movement of global capital. Atwood here emphasises not just the formidable power of biocapital but also the dangers posed by lack of regulation or even debate over what is legitimate in this field of research. As she has pointed out, there is nothing in this fictional text that has not already happened somewhere in the world, and the laboratories in her imagined future, where scientists have complete freedom to experiment, have their analogues in what Nicholas Agar calls the 'underground' research currently carried out in countries not bound by Western legislation over reproductive technologies.[19] He cites CLONAID, an organisation founded by a UFO cult that claims to have cloned human beings and has almost certainly experimented on human embryos. Agar argues that such underground research will allow us to pass through 'the ethically impossible passage to safe enhancement technologies' whether we like it or not, by which he means that unethical experiments entailing, for example, the cloning of human beings who will suffer from serious disabilities will be carried out regardless of Western concerns, because the demand is there (174). He argues that we should not discount the knowledge derived from such experimentation on moral grounds, as we might reject the data derived from Nazi medical experiments, for example: from his perspective, pioneers in human enhancement should be seen as closer to the early experimenters in vaccination or organ transplantation. In other words, the end justifies the means.

Atwood's novel contests such pragmatic acceptance of unethical, end-directed procedures. Moreover, it engages in depth with the underlying motives that drive the eugenic ambitions of Jimmy's friend Crake. Crake is a genius, a graduate of the Watson-Crick Institute, who rises to the top of the biotech industry. He is, however, a troubled soul, affected by his parents' divorce and the fact that his father was murdered when he was on the point of blowing the whistle on a scheme to engineer diseases in order to increase profit margins. Crake's analysis of the problems of humanity is that they are caused by 'war, which is to say misplaced sexual energy', a diagnosis that reflects his father's unhappy life together with his own loneliness and sexual frustration (345). With his exceptional abilities and unrivalled access to genetic technologies, Crake decides that he will solve these problems by replacing human beings with a new humanoid species, the Crakers. Their sexual life will be cyclical and communal, short-circuiting the gap between desire and fulfilment so that there will be 'no more unrequited love . . . no more thwarted lust; no more shadow between the desire and the act': no more frustration, no more war. However, in order to make a clean break with the unsatisfactory past, Crake first has to release a virus that wipes out humanity, freeing the Crakers to make a new world. Atwood here underscores the ambivalence of Crake's eugenic programme, in which destruction and creation are inextricably entwined. The text also probes the psychology behind the enduring human fascination with eugenics. Crake points out that human beings are unique in their preoccupation with the lives of their descendants and attributes this to the fact that they are the only species that can foresee their own death. It is for this reason, he argues, that they cling to the hope that 'they can stick their souls into someone else, some new version of themselves, and live on forever' (139). In designing the Crakers to be everything that he is not, Crake imagines he has avoided such a naïve projection of his own needs, although his concern to eliminate his own faults can be seen as an equally solipsistic project.

Nikolas Rose offers a very different perspective on such an investment in the future of others. In *The Politics of Life Itself*, he describes what he sees as a shift in Western democracies away from the traditional belief in 'immortality in a promised land' to a collective investment in a future 'in which we will live forever through our offspring'. This investment, he suggests, requires us 'to commit our own energies to the future biological health of those who will carry us forward into the future—to our children and their own healthy lives' (257–8). For Rose, such a commitment is an ethical imperative, a part of what he calls a 'somatic ethics', which is developing in response to advances in genetic technology. This somatic ethics is not the responsibility of the state but is being forged through interactions between individuals and 'professionals of vitality'—doctors, genetic counsellors, biotechnology executives, research scientists, employees of drugs companies and so on. Rose is optimistic about the outcome, arguing that '[t]he political vocation of the life sciences today is tied to the belief that in most, maybe all cases . . . the risky, damaged, defective, or afflicted individual, once identified and assessed, may be treated or transformed by medical intervention at the molecular level' (253–4). However, as Rose acknowledges, the knowledge that will be created through advances in biomedical science means that in a world of limited resources, we will be confronted by difficult decisions about the relative worth of different kinds of life. As Troy Duster has argued, we do not make such decisions in a vacuum: normative assumptions about lives worth living have a powerful influence on the patients and health professionals making these choices. It is such deeply embedded assumptions that have captured the attention of writers like Hoffman, Ishiguro and Atwood, who have illuminated the extent to which our culture remains haunted by what David Macey has called 'the phantasy of a pure body, be it the individual body or the body politic itself'.[20] These writers also offer a critical perspective on the investment in the future invoked by Rose, highlighting the dangers inherent in making decisions on behalf of future human beings. They examine the ways in which eugenic choices have the potential to distort the relationship between parents and children, and they assess the potential for the emergence of a biological underclass as a result of differential access to molecular medicine. Their novels can be seen as richly imagined case histories that extend our understanding of the issues surrounding developments in biomedicine, and they draw attention to the extension of biopower associated with them. In so doing, they alert us to the continuities and disjunctions between the old eugenics and the new, and they invite critical consideration of the implications of liberal eugenics in a global bioeconomy.

Notes

INTRODUCTION

1. Sir William Beveridge, 'Eugenic Aspects of Children's Allowances' in *Twelve Galton Lectures: A Centenary Selection with Commentaries*, ed. by Steve Jones and Milo Keynes (London: Galton Institute, 2007), p. 85.
2. Pauline Mazumdar, *Eugenics, Human Genetics and Human Failings: The Eugenics Society, Its Sources and Its Critics in Britain* (London: Routledge, 1992), p. 258.
3. Sir Francis Galton, *Inquiries into Human Faculty and Its Development* (London: Macmillan, 1883), p. 25.
4. Sir Francis Galton, *Essays in Eugenics* (London: Eugenics Education Society, 1909), p. 42.
5. Richard A. Soloway, *Demography and Degeneration: Eugenics and the Declining Birthrate in Twentieth-Century Britain* (Chapel Hill and London: University of North Carolina Press, 1995), p. 22.
6. Quoted in Soloway, p. 21.
7. Galton, *Essays in Eugenics*, p. 7.
8. Michel Foucault, *Society Must Be Defended: Lectures at the Collège de France, 1975–76,* trans. David Macey (London: Penguin, 2004), p. 243.
9. As an example of an earlier struggle of this kind, Foucault gives the conflict in England after the Norman Conquest, when the monarchy was opposed by a bourgeoisie that defined itself in terms of its Anglo-Saxon heritage.
10. Quoted in Faith Schenk and A. S. Parkes, 'The Activities of the Eugenics Society', *Eugenics Review* 60 (September 1968), 155.
11. Ordoliberalism was a form of liberal economic theory developed before the Second World War and associated with the journal *Ordo*.
12. Quoted in Michel Foucault, *The Birth of Biopolitics: Lectures at the Collège de France, 1978–1979* (Basingstoke: Palgrave Macmillan, 2008), p. 198. Subsequent references are incorporated into the text.
13. See Michel Foucault, *La Naissance de la Biopolitique* (Paris: Seuil/Gallimard, 2004), p. 222.
14. See Foucault, *Society Must Be Defended*, p. 250.
15. Foucault's argument chimes with Richard Titmuss's claim that the stresses of war made central government aware for the first time of the strategic necessity of having a population that was efficient and physically fit—although Titmuss also stressed the affective changes brought by shared wartime experience, which led to 'a deeper comprehension of social justice' (quoted in Pat Thane, *Foundations of the Welfare State*, 2nd edition [Harlow: Pearson Education, 1996], p. 211).

16. See Evelyn Fox Keller, *The Mirage of a Space between Nature and Nurture* (Durham: Duke University Press, 2010). See Chapter 1 for a discussion of this.
17. The term 'particulate inheritance' refers to the theory that characteristics can be passed from generation to generation through discrete particles that later became conceptualised as genes.
18. Sir Francis Galton, *English Men of Science: Their Nature and Nurture*, quoted in Fox Keller, p. 23.
19. Daniel J. Kevles, *In the Name of Eugenics: Genetics and the Uses of Human Heredity* (Cambridge, MA: Harvard University Press, 1995), p. 173.
20. Thane, p. 154. She refers here particularly to the march of unemployed men from Jarrow, in the north east of England, to London, to deliver a petition to Downing St.
21. Quoted in Kevles, p. 174.
22. Richard Titmuss, *Poverty and Population: A Factual Study of Contemporary Social Waste* (London: Macmillan & Co, 1938), p. 5.
23. See James Watson, *The Double Helix*, with a new introduction by Steve Jones (London: Penguin, 1999), p. 155, for an account of Francis Crick's first use of this phrase.
24. Lily E. Kay, *Who Wrote the Book of Life: A History of the Genetic Code* (Stanford, CA: Stanford University Press, 2000), p. 30. Crick later admitted that he did not understand the precise meaning of the term dogma: see Francis Crick, *What Mad Pursuit: A Personal View of Scientific Discovery* (New York: Basic Books, 1988), p. 109.
25. For a discussion of contemporary epigenetics, see John Marks, 'The New Eugenics: Jacques Testart and French Bioethics', *New Formations* 60 (Spring 2007), 132.
26. The standard opening print run for Pelican paperbacks was 50,000, and many sold far more. For example, C. H. Waddington's *The Scientific Attitude* sold 100,000 copies in 1941. Waddington Papers, Edinburgh University Library, MS3053 (letter to Jean-Jacques Salomon, 13 February 1967).

NOTES TO CHAPTER 1

1. Michael Young, *The Rise of the Meritocracy: An Essay on Education and Equality* (1958) (Harmondsworth: Penguin, 1961), p. 40.
2. Carolyn Steedman, *Landscape for a Good Woman: A Story of Two Lives* (London: Virago, 1986), p. 122.
3. Young made this point very clearly in an article published in 2001, writing that '[t]he book was a satire meant to be a warning (which needless to say has not been heeded) against what might happen to Britain between 1958 and the imagined final revolt against the meritocracy in 2033'. See Michael Young, 'Down with Meritocracy', *The Guardian* (29 June 2001), available at <http://www.guardian.co.uk/comment/story/> [accessed 2 August 2007].
4. The Norwood Report, *Curriculum and Examination in Secondary Schools*, quoted in Pat Thane, *Foundations of the Welfare State*, 2nd edition (Harlow: Pearson Education Ltd, 1996), p. 215.
5. See papers at the National Archives, Public Record Office, ED 12/479 Norwood Report: minutes, committee papers, replies to questionnaire, evidence. Norwood Committee 1941–3 (Secondary Schools Examination Council).
6. Raymond B. Cattell, *The Fight for Our National Intelligence*, with introductions by Lord Horder, Major Darwin and F. P. Armitage (London: P. S. King & Son, 1937), p. 5.

7. See the Royal Commission on Population Report (London: HMSO, 1949), p. 155.
8. For an account of this stage in Burt's career, see Daniel J. Kevles, *In the Name of Eugenics: Genetics and the Uses of Human Heredity*, with a new preface by the author (Cambridge, MA and London: Harvard University Press, 1996), p. 83.
9. The papers in question are: C. L. Burt, 'The Evidence for the Concept of Intelligence', *British Journal of Educational Psychology* XXV (1955), 158–77; C. L. Burt, 'The Inheritance of Mental Abilities', Bingham Memorial Lecture, *American Psychology* XlII (1958), 1–15; C. L. Burt, 'The Genetic Determination of Intelligence: A Study of Monozygotic Twins Reared Together and Apart', *British Journal of Psychology* LVlI (1966), 137–53. See L. S. Hearnshaw, *Cyril Burt: Psychologist* (Ithaca, NY: Cornell University Press, 1979) for a full discussion of this issue.
10. See Hearnshaw, p. 188.
11. Greta Jones, *Social Darwinism and English Thought: The Interaction between Biological and Social Theory* (Brighton: Harvester Press, 1980), p. 171.
12. Pauline Mazumdar, *Eugenics, Human Genetics and Human Failings: The Eugenics Society, Its Sources and Its Critics in Britain* (London: Routledge, 1992), p. 52.
13. Quoted in Thane, p. 189.
14. Michael Young, *The Rise of the Meritocracy, 1870–2033* (1958) (Harmondsworth: Penguin, 1963), p. 32.
15. C. P. Blacker, *Eugenics: Galton and After* (London: George Duckworth & Co. Ltd., 1952), p. 253.
16. Young refers here to H. J. Eysenck's *Uses and Abuses of Psychology* (Harmondsworth: Penguin, 1953).
17. Young was also connected with Richard Titmuss, one of most prominent advisors to successive post-war governments, through his work at the Institute of Community Studies.
18. Raymond Williams, *The Long Revolution* (1961) (London: Hogarth Press, 1992), p. 47.
19. Raymond Williams, *Culture and Society 1780–1950* (1958) (Harmondsworth: Penguin, 1962), p. 131.
20. Raymond Williams, *Border Country* (1960) (Cardigan: Parthian Library of Wales, 2006), p. 4.
21. Michel Foucault, *Society Must Be Defended: Lectures at the Collège de France 1975–76*, trans. David Macey (London: Penguin, 2004), p. 255.
22. Peter Shore, 'In the Room at the Top', in *Conviction*, ed. by Norman Mackenzie (London: MacGibbon & Kee, 1958), p. 51.
23. John Braine, *Room at the Top* (1957) (London: Arrow Books, 1997), p. 131.
24. See Pierre Bourdieu, 'Cultural Reproduction and Social Reproduction', in *Power and Ideology in Education*, ed. by J. Karabel and A. H. Halsey (Oxford and New York: Oxford University Press, 1977), pp. 487–511. Bourdieu suggests that culture shares many of the characteristics of economic capital and that cultural habits and dispositions act as a resource that can be monopolised by individuals and groups and passed down through the generations.
25. Geoffrey Gorer, *The Danger of Equality and Other Essays* (London: Cresset Press, 1966), p. 243.
26. Marilyn Strathern, *Reproducing the Future: Anthropology, Kinship and the New Reproductive Technologies* (Manchester: Manchester University Press, 1992), p. 31.

27. John Osborne, *Look Back in Anger* (1957; first performed at the Royal Court Theatre, London, May 1956) (London: Faber and Faber Ltd, 1983), pp. 37–8.
28. See, for example, Andrzej Gasiorek, *Post-war British Fiction: Realism and After* (London: Hodder, 1995); and more recently Katherine Cockin and Jago Morrison (eds), *The Post-war British Literature Handbook* (London: Continuum, 2009).
29. Doris Lessing, *Walking in the Shade: Volume Two of My Autobiography, 1949–1962* (1997) (London: Flamingo, 1998), p. 213.
30. Richard Hoggart's account of the conflicts experienced by the scholarship boy was particularly influential. He argued that they were 'at the friction-point of two cultures' and were 'emotionally uprooted from their class, often under the stimulus of a stronger critical intelligence or imagination, qualities which can lead them into an unusual self-consciousness'. See *The Uses of Literacy: Aspects of Working-Class Life with Special Reference to Publications and Entertainments* (1957) (Harmondsworth: Penguin, 1962), p. 292.
31. Richard Titmuss, *Essays on 'The Welfare State'* (London: George Allen and Unwin Ltd, 1958), p. 90.
32. Kevles discusses the Myrdals' eugenic beliefs in 'Eugenics and Human Rights', *British Medical Journal* 319 (14 August 1999), 435–8. He notes that Alva Myrdal spoke for her husband, Gunnar, and for numerous liberals like themselves when in 1941 she wrote, 'In our day of highly accelerated social reforms the need for sterilization on social grounds gains new momentum. Generous social reforms may facilitate home-making and childbearing more than before among the groups of less desirable as well as more desirable parents.'
33. For a discussion of the relationship between eugenics and feminism in the period 1880–1920, see Lesley A. Hall, 'Women, Feminism and Eugenics' in *Essays in the History of Eugenics*, ed. by Robert A. Peel (London: Galton Institute, 1998); Clare Hanson, *A Cultural History of Pregnancy: Pregnancy, Medicine and Culture, 1750–2000* (Basingstoke and New York: Palgrave Macmillan, 2004); Angelique Richardson, *Love and Eugenics in the Late Nineteenth Century: Rational Reproduction and the New Woman* (Oxford: Oxford University Press, 2003).
34. Alva Myrdal and Viola Klein, *Women's Two Roles: Home and Work* (1956), 2nd revised edition (London: Routledge and Kegan Paul Ltd, 1968), p. 116.
35. Bowlby advocated continuous maternal care for children. He argued for the prime importance of the mother–child bond in promoting healthy development, and he contended that there was a direct connection between 'maternal deprivation' and juvenile delinquency. His book *Childcare and the Growth of Love* (Harmondsworth: Penguin, 1953) was especially influential. However, as critics were keen to point out, his conclusions in this book were based on the study of children who had been completely deprived of maternal care in the circumstances of war-time evacuation. See, for example, S. Yudkin and A. Holme, *Working Mothers and Their Children* (London: Michael Joseph, 1963).
36. Hilary Rose, 'Eugenics and Genetics: The Conjoint Twins?', *New Formations* 60 (Spring 2007), p. 15.
37. C. P. Blacker, *Eugenics: Galton and After*, p. 306.
38. C. O. Carter, 'Promising Families: Some Conclusions', *Eugenics Review* 52 (1961), 197.
39. See Betty Friedan, *The Feminine Mystique* (Harmondsworth: Penguin, 1963) for a discussion of 'the problem with no name'.

40. Judith Hubback, *Wives Who Went to College* (London: Heinemann, 1957), p. 3.
41. Penelope Mortimer, *The Pumpkin Eater* (1962) (Harmondsworth: Penguin, 1964), p. 43.
42. Hannah Gavron, *The Captive Wife: Conflicts of Housebound Mothers* (London: Routledge and Kegan Paul Ltd, 1966), p. 143 (my emphasis).
43. Margaret Drabble, *Jerusalem the Golden* (1967) (Harmondsworth: Penguin, 1972), p. 82.
44. Carolyn Steedman, *Landscape for a Good Woman: A Story of Two Lives* (London: Virago, 1986), p. 15.
45. See Nancy Chodorow, *The Reproduction of Mothering: Psychoanalysis and the Sociology of Gender* (Berkeley: University of California Press, 1979).
46. The paediatrician and psychoanalyst D. W. Winnicott developed his influential concept of the 'good enough mother' in *The Child, the Family and the Outside World* (Harmondsworth: Penguin, 1964).
47. Lorna Sage, *Bad Blood* (2000) (London: Fourth Estate, 2001), pp. 161–2.
48. See Denis Wheatley, *To the Devil, a Daughter* (London: Hutchinson, 1953).
49. Hilary Land, 'We Sat Down at the Table of Privilege and Complained About the Food', in *The Rise and Rise of Meritocracy*, ed. by Geoff Dench (Oxford: Blackwell, 2006), p. 56.
50. Sheila Rowbotham, *Threads through Time: Writings on History and Autobiography* (Harmondsworth: Penguin, 1999), p. 77.
51. See Anthony Heath and Clive Payne, 'Twentieth Century Trends in Social Mobility in Britain', Centre for Research into Elections and Social Trends, Working Paper No. 70, June 1999. This study is the first to examine the social mobility of women and men in the post-war period. An extensive earlier study based on a 1972 survey excluded women from consideration because of the methodological difficulties involved in determining their class position. See John H. Goldthorpe, with Catriona Llewellyn and Clive Payne, *Social Mobility and Class Structure in Modern Britain*, 2nd edition (Oxford: Clarendon Press, 1980), pp. 277–301 for a discussion of this issue. Debate continues over whether the father's or husband's position can be assumed to be the best guide to the social class of the household as a whole, particularly in view of the growing number of female-headed households and dual-career families.
52. Geoff Dench, 'Introduction: Reviewing Meritocracy', in Dench, p. 8.
53. Michael Young, 'Looking Back on *Meritocracy*', in Dench, p. 77.

NOTES TO CHAPTER 2

1. David Stafford-Clark, *Psychiatry Today* (London: Penguin, 1961), p. 90.
2. Joanna Ryan with Frank Thomas, *The Politics of Mental Handicap* (Harmondsworth: Penguin, 1980), p. 15.
3. The terms of the legislation are quoted in Kathleen Jones, *A History of the Mental Health Services* (London and Boston: Routledge and Kegan Paul, 1972), pp. 204–5.
4. The number of patients receiving out-patient care rose from 83,984 to 97,732 in 1970, probably due to an increase in the provision of these services (see Jones, pp. 362–3).
5. The percentage of the population with learning difficulties is currently estimated to be between 1 and 2%. Exact statistics are difficult to obtain; these figures are taken from the British Institute of Learning Disabilities.

6. Marshall was a distinguished sociologist who specialised in educational policy. He gave a lecture to the Eugenics Society in 1953 on 'Social Selection and the Welfare State', which pointed out that grammar school places were being allocated disproportionately to the middle classes.
7. C. P. Blacker, *Neurosis and the Mental Health Services* (first published 1946) (London, New York, Toronto: Oxford University Press, 1948, p. 48), my emphasis.
8. Stafford-Clark was not known for his progressive views. He invoked the spectre of animality in his discussion of mental deficiency in *Psychiatry Today*, and the textbook he co-authored (*Psychiatry for Students*, with A. Smith) provoked widespread protests to the publishers in connection with its treatment of mental handicap. See Ryan and Thomas for a discussion of this.
9. Michel Foucault, *Society Must Be Defended: Lectures at the Collège de France, 1975–76*, trans. David Macey (London: Penguin, 2004), p. 255.
10. Mathew Thomson, *The Problem of Mental Deficiency: Eugenics, Democracy, and Social Policy in Britain c.1870–1959* (Oxford: Clarendon Press, 1998), p. 282.
11. Nikolas Rose, 'The Politics of Life Itself', *Theory, Culture and Society* 18 (2001), 4.
12. See Pauline Morris, *Put Away: A Sociological Study of Institutions for the Mentally Retarded*, with a foreword by Peter Townsend (London: Routledge and Kegan Paul, 1969), p. 21. Morris quotes from an unpublished paper by one consultant alleging that 'sexual problems loom high in the lives of these patients; homosexuality, child assault, rape, exposure with the men and prostitution and some perversion with the women'. Such prejudice was unusual among specialists in the field, but the sexual segregation of mental deficiency patients, perhaps in response to fears among the wider public, was real enough.
13. See Pauline Mazumdar, *Eugenics, Human Genetics and Human Failings: The Eugenics Society, Its Sources and Its Critics in Britain* (London and New York: Routledge, 1992), chapter 5 for a discussion of the 'social problem group'.
14. C. P. Blacker (ed), *Problem Families: Five Inquiries* (London: Eugenics Society, 1952), p. 14.
15. Shelagh Delaney, *A Taste of Honey* (London: Methuen, 1963), p. 43. The play was first performed at the Theatre Royal, Stratford East in May 1958. It was made into an award-winning film directed by Tony Richardson in 1961.
16. Peter Nichols, *A Day in the Death of Joe Egg* (London: Faber and Faber, 1967), p. 63. The play was first performed at the Citizen's Theatre in Glasgow in May 1967.
17. There is a long tradition of connecting birth defects with the experience of the mother. The best-known is the belief that a cleft palate (or 'hare lip') could be caused by the sight of a hare in pregnancy.
18. Eva M. Hubback, *The Population of Britain* (Harmondsworth: Penguin, 1947), p. 237.
19. D. R. Mace, *Marriage Counselling: The First Full Account of the Remedial Work of the Marriage Guidance Counsellors* (London: J. & A. Churchill Ltd, 1948), p. 144.
20. See the papers of C. P. Blacker, Wellcome Library, for the letter from Margery Spring Rice and a copy of the pamphlet 'Eugenic Guidance'.
21. Quoted in Blacker, *Problem Families*, p. 15.
22. See Gillian Swanson, 'Serenity, Self-regard and the Genetic Sequence: Social Psychiatry and Preventive Eugenics in Britain, 1930s–1950s', *New Formations* 60 (Spring 2007), 50–65.

23. Quoted by Swanson, p. 63.
24. This accords with the fact that in many of the 'case histories' in *Problem Families*, it is noted that despite the conditions in which they lived, the majority of the children were happy. See for example the account of the 'O' family; although 'the home has for years been in a state of destitution . . . [t]he children appear happy' (100).
25. L. T. Hilliard and Brian Kirman, *Mental Deficiency*, 2nd edition (London: J. & A. Churchill Ltd, 1965), p. 39.
26. L. T. Hilliard, 'Resettling Mental Defectives: Psychological and Social Aspects', *British Medical Journal* 1 (1954), 1372–4.
27. Hilliard and Kirman, *Mental Deficiency*, p. 56.
28. These included Cyril Burt and J. B. S. Haldane.
29. This point is made by Alison Light in *Forever England: Femininity, Literature and Conservatism between the Wars* (London and New York: Routledge, 1991), p. 77.
30. See Light, pp. 105–6.
31. Margery Allingham, *The China Governess* (1963) (London: Penguin, 1965), p. 11. Allingham also explores the theme of mental deficiency in *The Mind Readers* (1965), in which a clever schoolboy invents a telepathic device that is tested out on a pair of working-class twins. One of them is a 'sub' (i.e. he is mentally defective) who commits suicide because of the strain of the experiments. Limited textual space is given to this catastrophe, which Allingham's aristocratic detective Campion takes coolly in his stride.
32. Blacker, *Problem Families*, p. 17.
33. Julia Kristeva, 'Approaching Abjection', excerpt from *Powers of Horror, The Portable Kristeva*, ed. by Kelly Oliver (Chichester and New York: Columbia University Press, 1997), p. 231.
34. See Mary Douglas, *Purity and Danger* (1966) (London: Routledge, 2002), pp. 47–50.
35. See Erving Goffman, *Asylums: Essays on the Social Situation of Mental Patients and Other Inmates* (Penguin Social Studies) (London: Penguin, 1991).
36. Morris, p. xxii.
37. Aubrey Lewis, 'Psychopathic Personality: A Most Elusive Category', *Psychological Medicine* 4 (1974), 137.
38. Hilliard and Kirman, *Mental Deficiency*, p. 52.
39. *A Textbook of Mental Deficiency*, first compiled by the late A. F. Tredgold, 10th edition by R. F. Tredgold and K. Soddy (London: Baillière, Tindall & Cox, 1963). In this text, some of the case histories originally compiled by A. F. Tredgold are recycled by his son, perpetuating outdated assumptions and prejudices.
40. Bentley was granted a posthumous royal pardon in 1993, and his conviction for murder was quashed in 1998.
41. Bill Osbergy, *Youth in Britain Since 1945* (Making Contemporary Britain) (Oxford: Wiley-Blackwell, 1997), p. 10.
42. See John Davis, *Youth and the Condition of Britain: Images of Adolescent Conflict* (London: Athlone Press, 1990) and Peter Lewis, *The Fifties* (London: Heinemann, 1978), quoted in Dominic Sandbrook, *Never Had It So Good: A History of Britain from Suez to the Beatles* (London: Little, Brown, 2005), p. 418. As Sandbrook points out, the rise in convictions did not mean that there was an overwhelming number of delinquent teenagers, but the increase was significant.
43. David Yallop, *To Encourage the Others* (London: W. H. Allen, 1971), p. 15.

44. Anthony Burgess, *A Clockwork Orange* (London: Penguin Classics, 2000), p. 42.
45. See, for example, Blake Morrison, 'Introduction' to *A Clockwork Orange*, p. xxii.
46. Peter Medawar, *The Future of Man* (New York: Mentor Books, 1961), p. 72. Medawar argues that it is likely that 'people of mediocre or rather lowly intelligence are intrinsically more fertile' than those with a very high or low IQ and that 'if innately unintelligent people tend to have larger families, then, with some qualifications, we can infer that the average level of intelligence may decline' (79–83).
47. Huxley had joined the Eugenics Society in the 1930s and was its president at this time.
48. Julian Huxley, *Essays of a Humanist* (Harmondsworth: Penguin, 1966), p. 272.
49. Julian Huxley (ed.), *The Humanist Frame* (London: Allen and Unwin, 1961), p. 19.
50. Julian Huxley, *New Bottles for New Wine* (London: Chatto & Windus, 1957), p. 17.
51. For a discussion of transhumanism, see Nicholas Agar, *Liberal Eugenics: In Defence of Human Enhancement* (Oxford and Victoria: Blackwell, 2004), pp. 16–9.
52. Francis Fukuyama, *Our Posthuman Future: Consequences of the Biotechnology Revolution* (New York: Farrar, Straus & Giroux, 2002), quoted in Agar, p. 133.
53. Roger Luckhurst, *'The Angle between Two Walls': The Fiction of J. G. Ballard* (Liverpool: Liverpool University Press, 1997), p. 37.
54. J. G. Ballard, *The Drought* (London: Flamingo, 2002), p. 3.
55. See David Bradshaw, 'Introduction' to Aldous Huxley, *Brave New World* (London: Flamingo, 1994) for a discussion of this.
56. See the first chapter of *Brave New World* for a description of babies in bottles.
57. J. G. Ballard, *Miracles of Life: Shanghai to Shepperton* (London: Fourth Estate, 2008), p. 137.
58. Doris Lessing, *The Four-Gated City: Volume Five of Children of Violence* (London: MacGibbon & Kee, 1969), p. 700.
59. See Goffman, *Asylums*; and for the anti-psychiatry movement R. D. Laing, *The Divided Self: An Existential Study in Sanity and Madness* (London: Penguin, 1990).
60. Jones, p. 337.
61. Morris, p. 281.
62. Ryan and Thomas, *The Politics of Mental Handicap.*
63. *The Warnock Report: Special Educational Needs* (London: HMSO, 1978).
64. Mary Warnock, *Special Educational Needs: A New Look* (London: Philosophy of Education Society of Great Britain, 2005).
65. See *Deinstitutionalisation and People with Intellectual Disabilities: In and Out of Institutions*, ed. by Kelley Johnson and Rannveig Traustadóttir (London and Philadelphia: Jessica Kingsley Publishers, 2005).

NOTES TO CHAPTER 3

1. Wilhelm Johannsen, 'The Genotypic Conception of Heredity', cited in Evelyn Fox Keller, *The Century of the Gene* (Cambridge, MA and London: Harvard University Press), p. 2.

2. C. D. Darlington, *Genetics and Man* (Harmondsworth: Penguin, 1964), p. 387.
3. Francis Crick, 'Discussion: Eugenics and Genetics' in *Man and His Future: A Ciba Foundation Volume*, ed. by Gordon Wolstenholme (Boston: Little, Brown & Co., 1963), p. 294.
4. Daniel J. Kevles, *In the Name of Eugenics: Genetics and the Uses of Human Heredity* (Cambridge, MA, and London: Harvard University Press, 1995), pp. 251–2.
5. Richard A. Soloway, *Demography and Degeneration: Eugenics and the Declining Birthrate in Twentieth-Century Britain* (Chapel Hill: University of North Carolina Press, 1995), p. 353.
6. For an account of this, see Kevles, pp. 251–2.
7. See, for example, Julian Huxley (ed.), *The Humanist Frame* (London: Allen & Unwin, 1961) and *Essays of a Humanist* (Harmondsworth: Penguin, 1964).
8. Crick, 'Discussion', p. 294.
9. See, for example, Watson's essay in 'Genetics and Eugenics: A Personal Odyssey', in *Medicine after the Holocaust: From the Master Race to the Human Genome*, ed. by Sheldon Rubenfeld in conjunction with the Holocaust Museum, Houston (Basingstoke: Palgrave, 2010).
10. Stefan Collini, 'Introduction' to C. P. Snow, *The Two Cultures* (Cambridge: Cambridge University Press, 1993), p. xxiii.
11. See James D. Watson, *The Double Helix* (London: Penguin, 1999), p. 25.
12. Doris Lessing, *Walking in the Shade: Volume Two of My Autobiography, 1949–1962* (London: Flamingo, 1998), p. 104.
13. Michel Foucault, 'Truth and Power', in *The Foucault Reader*, ed. by Paul Rabinow (Harmondsworth: Penguin, 1986), p. 51. Foucault refers to this as a 'sordid affair' because of the part played by the French Communist Party in suppressing criticism of Stalin's regime.
14. For a detailed discussion of the relationship between Lysenko's ideas and Lamarckian thought, see Dominique Lecourt, *Proletarian Science?: The Case of Lysenko*, with an introduction by Louis Althusser, trans. Ben Brewster (London: New Left Books, 1977), pp. 97–8.
15. Cited in Lecourt, p. 96. When the technique of close planting was used on a large scale in Stalin's 'Great Plan for the Transformation of Nature' in 1949, it had disastrous results.
16. See Daniel Todes, 'Global Darwin: Contempt for Competition', *Nature* 7269 (November 2009), 36–7.
17. Nils Roll-Hansen, 'Lamarckism and Lysenkoism Revisited' in *Transformations of Lamarckism*, ed. by Snait B. Gissis and Eva Jablonka (Cambridge: MIT Press, 2011), p. 81.
18. For details of other victims of the purges, see Diane C. Paul, 'A War on Two Fronts: J. B. S. Haldane and the Response to Lysenkoism in Britain', *Journal of the History of Biology* 16 (Spring 1983), 8–9.
19. Julian Huxley, 'Soviet Genetics: The Real Issue', *Nature* 163 (June 1949), 936.
20. The term 'Modern Synthesis' describes the fusion of evolutionary theory with genetic science, which explained the mechanisms of inheritance, thus fleshing out Darwin's theory. Julian Huxley's *Evolution: The Modern Synthesis* (London: George Allen and Unwin, 1942) popularised the term.
21. Darlington's comments are quoted in Oren Solomon Harman, *The Man Who Invented the Chromosome: A Life of Cyril Darlington* (Cambridge, MA, and London: Harvard University Press, 2004), p. 158.
22. See Paul, 14–15.

23. C. H. Waddington, 'Lysenko and the Scientists', *The New Statesman and Nation* (25 December 1948), 566.
24. John Wyndham, *The Day of the Triffids* (London: Penguin Classics, 1999), p. 23.
25. See Evelyn Fox Keller, *Making Sense of Life: Explaining Biological Development with Models, Metaphors and Machines* (Cambridge, MA, and London: Harvard University Press, 2002).
26. Marion J. Lamb, 'Attitudes to Soft Inheritance in Great Britain, 1930s–1970s', in Gissis and Jablonka, p. 117.
27. C. D. Darlington, *Genetics and Man* (Harmondsworth: Penguin, 1964), p. 348.
28. Julian Huxley, *Essays of a Humanist*, p. 269.
29. Aldous Huxley, *Brave New World Revisited* (London: Vintage, 2004), pp. 21–2.
30. This correspondence is in the Eugenics Society Archive, Wellcome Library (SA/EUG/C/186).
31. See Kevles, p. 262.
32. Julian Huxley, 'Eugenics in Evolutionary Perspective' in *Essays of a Humanist*, p. 275.
33. John Wyndham, *The Midwich Cuckoos* (London: Penguin, 2008), p. 123.
34. Ernst Mayr, 'Change in Genetic Environment and Evolution' in *Evolution as a Process*, ed. by Julian Huxley, A. C. Hardy and E. B. Ford (London: George Allen and Unwin, 1954), pp. 157–80.
35. Quoted in *Report on the Effect of Radiation on Human Heredity* (Geneva: World Health Organization, 1957), p. 12.
36. Peter Alexander, *Atomic Radiation and Life* (Harmondsworth: Penguin, 1957), p. 129 (my emphasis).
37. David Seed, *American Science Fiction: Literature and Film* (Edinburgh: Edinburgh University Press, 1999), p. 55.
38. Michel Foucault, *Society Must Be Defended: Lectures at the Collège de France 1975–76*, trans. David Macey (London: Penguin, 2004), p. 253.
39. John Wyndham, *The Chrysalids* (London: Penguin, 1958), p. 18.
40. Foucault, *Society Must Be Defended*, p. 252.
41. Ibid., p. 259.
42. Doris Lessing, 'Learning to Put the Questions Differently' in *Putting the Questions Differently: Interviews with Doris Lessing*, ed. by Earl G. Ingersoll (London: Flamingo, 1996), pp. 25–6.
43. Lessing, *Walking in the Shade*, p. 113.
44. Ibid., p. 111.
45. Martha Quest is the main protagonist in all five novels in Lessing's *Children of Violence* series. *The Four-Gated City* is the concluding volume.
46. R. D. Laing's *The Divided Self* (1960) was the founding text of the anti-psychiatry movement in the UK. Lessing knew Laing, and her lover Clancy Sigal was part of Laing's therapeutic community at Kingsley Hall in the East End of London.
47. Doris Lessing, *The Four-Gated City: Volume Five of the Children of Violence* (London: MacGibbon and Kee, 1969), p. 553.
48. Doris Lessing, *Re: Colonised Planet 5 Shikasta* (1979) (London and New York: Granada, 1981), p. 15.
49. Lessing has written at length about the influence of Idries Shah's teachings. See, for example, 'In the World, Not of It', *Encounter* (August 1972), 61–64; 'If You Knew Sufi', *Guardian* (8 January 1975), 12.
50. *Stranger in a Strange Land* (1961) is a science fiction classic by Robert A. Heinlein, in which the earth is transformed by a Martian invader, in a process that has been seen as paradigmatic of colonial encounters.

51. Doris Lessing, 'Preface' to *The Sirian Experiments* (London: Flamingo, 1994), p. 35.
52. Francis Galton, *Hereditary Genius*, 2nd edition (1892) (London: Fontana, 1962), p. 45.
53. Naomi Mitchison, *All Change Here: Girlhood and Marriage* (London: Bodley Head, 1975), p. 61.
54. McLaren was a leading developmental biologist. Her pioneering work on embryo transfer contributed significantly to techniques in assisted reproduction. She played an important part in the discussion of ethical issues relating to the use of embryos in research and was a key member of the committee on human fertilisation chaired by Mary Warnock. She subsequently served as a member of the Human Fertilisation and Embryology Authority (HFEA).
55. Anne McLaren, 'Dealing with Life' in *What the Human Race Is Up To*, ed. by Naomi Mitchison (London: Gollancz, 1962), p. 171.
56. Susan Merrill Squier, *Babies in Bottles: Twentieth Century Visions of Reproductive Technology* (New Brunswick, NJ: Rutgers University Press, 1994), pp. 186, 188.
57. Naomi Mitchison, *Memoirs of a Spacewoman* (1962) (London: New English Library, 1976), p. 53.
58. See Peter Gluckman and Mark Hanson, *Mismatch* (Oxford: Oxford University Press, 2006) for a summary of recent work in this area.
59. See Evelyn Fox Keller, *The Century of the Gene* (Cambridge, MA, and London: Harvard University Press, 2000).
60. See Hilary Rose, 'Eugenics and Genetics: The Conjoint Twins?', *New Formations* 60 (Spring 2007), 13–26.

NOTES TO CHAPTER 4

1. H. J. Muller, 'Comments and Criticisms on Different Items of the Statement' in *The Race Concept: Results of an Inquiry* (Paris: UNESCO, 1952), p. 52.
2. G. C. L. Bertram, *West Indian Immigration* (London: Eugenics Society, 1958), p. 17.
3. Francis Crick, letter to John T. Edsell, 19 June 1971 (Crick Papers, Wellcome Library, PP/CRI/D2/14).
4. The biological category of race has now been thoroughly discredited; for an excellent discussion of this, see, among others, Richard S. Cooper, M.D., Jay. S. Kaufman, Ph.D., and Ryk Ward, Ph.D., 'Race and Genomics', *New England Journal of Medicine* 348 (20 March 2003), 1166–70. Scholars sometimes use inverted commas around the term 'race' to signal their recognition of this fact. However, in this chapter, I use the word without quotation marks to reflect common usage in the period under discussion.
5. Michel Foucault, *Society Must Be Defended: Lectures at the Collège de France, 1975–76*, trans. David Macey (London: Penguin, 2004), pp. 256–7.
6. Francis Galton, *Hereditary Genius: An Inquiry into Its Laws and Consequences* (London: Macmillan, 1869). In his chapter on 'The Comparative Worth of Different Races', Galton argues that the ancient Greeks were the most highly developed of all 'races' before they succumbed to decadence and degeneration, while African and American Negroes are as far below the Anglo-Saxon race as the Anglo-Saxons are below the ancient Greeks. His description of 'merciful eugenics' appears in *Inquiries into Human Faculty and Its Development*, cited in Laura Otis (ed.), *Literature and Science in the Nineteenth Century: An Anthology* (Oxford: Oxford World's Classics, 2002), p. 479.

7. *UNESCO and Its Programme, Vol. 3 The Race Question* (Paris: UNESCO, 1950), p. 1.
8. The term 'race science' (or 'scientific racism') refers to an understanding of race as grounded in anatomical and physiological differences, which could be scientifically verified. This approach was characteristic of physical anthropology in the nineteenth and early twentieth centuries.
9. The comments from the Royal Anthropological Society are cited in Michelle Brattain, 'Race, Racism and Antiracism: UNESCO and the Politics of Presenting Science to the Postwar Public', *The American Historical Review* 112 (2007), available at <http://www.historycooperative.org/journals/ahr/112.5/brattain.html> [accessed 15 June 2011]. In this illuminating article, Brattain also explores the influence of the progressive social anthropologist Ashley Montagu on the 1950 statement.
10. L. C. Dunn, *Report on Meeting of Physical Anthropologists and Geneticists for a Definition of the Concept of Race* (Paris: UNESCO, 1951), p. 2.
11. *The Race Concept*, p. 13.
12. C. D. Darlington, *The Facts of Life* (London: George Allen & Unwin, 1953), pp. 292–3.
13. For a discussion of 'race crossing', see Lucy Bland, 'British Eugenics and "Race Crossing": A Study of an Interwar Investigation', *New Formations* 60 (Spring 2007), 66–78. For the mobilisation of hybridity as a cultural (as opposed to a biological) concept, see Homi Bhabha, 'Signs Taken for Wonders' in *The Location of Culture* (London and New York: Routledge, 1994); Salman Rushdie, *Imaginary Homelands* (New York and London: Penguin, 1991).
14. C. D. Darlington, 'The Genetic Understanding of Race in Man', *International Social Science Bulletin* 2.4 (Winter 1950), 487.
15. *The Facts of Life*, p. 357. This passage is reprinted in Darlington's *Genetics and Man* (Harmondsworth: Penguin, 1966).
16. For an account of Blacker's introduction of the term 'crypto–eugenics', see Faith Schenk and A. S. Parkes, 'The Activities of the Eugenics Society', *Eugenics Review* 60 (September 1960), 154.
17. These figures are taken from Charles Husband (ed.), *'Race' in Britain: Continuity and Change* (London: Hutchinson, 1982), p. 83.
18. G. C. L. Bertram, *West Indian Immigration* (London: Eugenics Society, 1958), p. 17.
19. Colin Bertram, *Adam's Brood: Hopes and Fears of a Biologist*, foreword by Sir Julian Huxley, F. R. S. (London: Peter Davies, 1959), p. 201.
20. See, for example, Bland, 'British Eugenics and "Race Crossing"' for a discussion of the ways in which the study of 'race-crossing' drew on the anthropometrical work of Karl Pearson.
21. The survey was carried out by Sheila Webster and is discussed in Michael Banton, *White and Coloured: The Behaviour of British People towards Coloured Immigrants* (London: Jonathan Cape, 1959). Neither Webster nor Banton considers the extent to which the responses were shaped by the form of the questions and the context in which they were asked.
22. George Lamming, *The Emigrants* (1954) (London and New York: Allison and Busby, 1980), p. 67.
23. Chris Waters, '"Dark Strangers" in Our Midst: Discourses of Race and Nation in Britain, 1947–1963', *The Journal of British Studies* 36 (April 1997), 210.
24. Sheila Patterson, *Dark Strangers: A Study of West Indians in London* (Harmondsworth: Penguin, 1965), p. 203.
25. Banton, p. 128.

26. See, for example, Steven Connor, *The English Novel in History: 1950 to the Present* (London and New York: Routledge, 1995), p. 91 for a positive response to MacInnes's writing.
27. *The Colin MacInnes Omnibus* (London: Allison and Busby, 2005), p. 161.
28. For other examples of the trope of the 'savage' dancing, see Sam Selvon, *The Lonely Londoners* (1956) and Muriel Spark, *The Ballad of Peckham Rye* (1960).
29. Patterson, p. 26.
30. Muriel Spark, *The Go-Away Bird and Other Stories* (Harmondsworth: Penguin, 1963), p. 15.
31. Frantz Fanon, *Black Skin, White Masks*, trans. Charles Lam Markmann (London: Pluto Press, 1986), p. 13.
32. Anna Marie Smith, *New Right Discourse on Race and Sexuality: Britain, 1968–1990* (Cambridge: Cambridge University Press, 1994), p. 146.
33. For details of the changes made by the Kenyan government, see Smith, p. 151.
34. Enoch Powell, 'Rivers of Blood' speech, reprinted in *The Daily Telegraph*, available at <http://www.telegraph.co.uk/comment/3643823/Enoch-Powells-Rivers-of-Blood-speech.html> [accessed 7 April 2011]. See also J. Enoch Powell and Rex Collings (eds.), *Reflections: The Selected Writings and Speeches of Enoch Powell* (London: Bellew Publishing Co. Ltd, 1992).
35. Speech of 16 November 1968, cited in Smith, p. 158.
36. See Smith, pp. 160, 162.
37. Christopher Priest, *Fugue for a Darkening Island* (London: Faber and Faber, 1972), p. 23.
38. Arthur S. Jensen, 'How Much Can We Boost IQ and Scholastic Achievement?', *Harvard Educational Review* 33 (1969), 1–123.
39. See Daniel J. Kevles, *In the Name of Eugenics: Genetics and the Uses of Human Heredity* (Cambridge, MA, and London: Harvard University Press, 1995), pp. 281–2 for a detailed account of Lewontin's arguments.
40. Cited in Kevles, p. 282.
41. See Steven Rose, 'Hans Eysenck's Controversial Career', *The Lancet* 376 (7 August 2010), 407–8 for an account of the response to Jensen and Eysenck in the UK. As Rose notes, he was closely involved with both the Jensen debate and the NUT pamphlet.
42. Hans Eysenck, *The IQ Argument: Race, Intelligence and Education* (New York: Library Press, 1971), p. 21.
43. Cited in Hazel V. Carby, '"Schooling in Babylon" in the Centre for Contemporary Cultural Studies' in *The Empire Strikes Back: Race and Racism in 70s Britain* (Abingdon and New York: Routledge, 1994), p. 187.
44. Francis Crick, letters to Peter Medawar, 31 January 1977, 17 February 1977 (Crick Papers, Wellcome Library, PP/CRI/D2/14).
45. Francis Crick, letter to John T. Edsell, 19 June 1971 (Crick Papers, Wellcome Library, PP/CRI/D2/14).
46. See Crick Papers, Wellcome Library, PP/CRI/D2/14 for this correspondence.
47. 'Resolution', *American Psychologist* 27 (July 1972), 660–1.
48. Steven Rose, review of Arthur Jensen, *Educability and Group Differences*, *New Scientist* (14 June 1973), 704–5.
49. Buchi Emecheta, *In the Ditch* (London: Flamingo, 1988), p. 30.
50. Cited in Errol Lawrence, 'In the Abundance of Water the Fool Is Thirsty: Sociology and Black "Pathology"' in *The Empire Strikes Back*, p. 117.
51. Banton, p. 122.
52. Carby, p. 190.
53. Patterson, p. 263 (my emphasis).

54. Cited in Lawrence, p. 117.
55. Carby notes that the Select Committee on Race Relations, 1968–69 expressed concern about the fact that 'many West Indian mothers work during school-hours and cannot get to school meetings' (190).
56. See Lawrence for an incisive critique of the concept of 'maladjustment' as it is used in the discourse of race relations (120).
57. This leaflet is in the Family Planning Association Papers, Wellcome Library, SA/FPA.
58. Details are in the FPA Papers, SA/FPA.
59. Jean Robinson's notes on this are in the Patients Association Papers, Wellcome Library, SA/PAT/D/14.
60. See the minutes of the first meeting of the Co-ordinating Group on Depo-Provera, also in the Patients Association Papers, Wellcome Library, SA/PAT/D/14.
61. *A Report by the Campaign against Depo-Provera* (1983), p. 23.
62. Beverley Bryan, Stella Dadzie and Suzanne Scafe, *The Heart of the Race: Black Women's Lives in Britain* (London: Virago, 1985), p. 105.
63. Emecheta, *In the Ditch*, pp. 140–1.
64. Buchi Emecheta, *Second-Class-Citizen* (London: Heinemann, 1994), pp. 152–3.
65. Powell used this phrase in a speech of 13 June 1970, cited in Smith, p. 163.

NOTES TO CHAPTER 5

1. Margaret Sanger, 'The Humanity of Family Planning' in *The Third International Conference on Planned Parenthood: Report of the Proceedings* (Bombay: Family Planning Association of India, 1952), p. 55.
2. Pandit Jawaharlal Nehru, Prime Minister, 'Message of Good Wishes', in *The Third International Conference*, 2.
3. Kingsley Davis, 'Population Policy: Will Current Programs Succeed?', *Science* 158 (10 November 1967), 738.
4. International Planned Parenthood Federation home page, available at <http://www.ippf.org/en/> [accessed 15 June 2011].
5. Karl Ittmann, 'Demography as Policy Science in the British Empire, 1918–1969', *Journal of Policy History* 15 (2003), 426.
6. Royal Commission on Population, *Report* (London: HMSO, 1949), p.134.
7. See Kathleen Paul, *Whitewashing Britain: Race and Citizenship in the Post-war Era* (Ithaca, NY, and London: Cornell University Press, 1997), p. 46 for a discussion of this.
8. Details of this funding are contained in the International Planned Parenthood Federation's second annual report. The IPPF head office was provided rent free by the Eugenics Society, and in the year ending 1954, the Eugenics Society gave a grant of £250, British Family Planning Associations contributed £371. The Brush Foundation for Race Betterment, run by Dorothy Brush gave £1064. IPPF papers, Wellcome Library, PP/MCS/B.9.
9. Susan Greenhalgh, 'The Social Construction of Population Science: An Intellectual, Institutional, and Political History of Twentieth-Century Demography', *Comparative Studies in Society and History* 38 (January 1996), 27.
10. Blacker discusses 'crypto eugenics' in a letter to Dorothy Brush, 13 April 1956 (papers of C. P. Blacker, Wellcome Library, PP.CPB/C.4/4). He expresses his beliefs about race in his book *Eugenics: Galton and After* (London: Duckworth, 1952), which was designed to re-frame eugenics to fit the post-war world. In an appendix discussing Galton's views on race, Blacker endorses

the view that not all human races have an equal power of 'sustaining and promoting the process known as Western civilization' (328).

11. C. P. Blacker, 'Voluntary Sterilization: Transitions throughout the World', *Eugenics Review* 54 (October 1962), 160.
12. Matthew Connelly, *Fatal Misconception: The Struggle to Control World Population* (Cambridge, MA, and London: Harvard University Press, 2008), p. 185.
13. Quoted in Connelly, p. 188 (emphasis in the original).
14. See Connelly, p. 311, for a discussion of Deverell's management style at the IPPF head office. It was considered to be 'an extension not only of British philosophy but virtually of that of the British Civil Service or Colonial Office'.
15. Quoted in Connelly, p. 217.
16. International Planned Parenthood Federation Leaflet, c. 1958, distributed worldwide (papers of C. P. Blacker, Wellcome Library, PP/CPB/C.4).
17. Julian Huxley, 'World Population', *Scientific American* 194 (1956), 69.
18. Paul Ehrlich, *The Population Bomb* (Cutchogue, NY: Buccaneer Books, 1971), p. 1. As Ehrlich notes, the title first appeared on a pamphlet published by the Hugh Moore Fund, and annual editions of the pamphlet had been widely distributed, totalling more than two million copies.
19. Aldous Huxley, 'The Double Crisis', *The Science News-Letter* 55 (26 March 1949), 200.
20. Aldous Huxley, *Island* (London: Vintage, 2004), p. 151.
21. Quoted in Connelly, p. 118.
22. Blacker made this case in an article titled '"Eugenic" Experiments Conducted by the Nazis on Human Subjects', *Eugenics Review* 44 (April 1952), in which he drew on the evidence he submitted to a government committee investigating the war crimes of German doctors. In the article, he argues that eugenics is 'in essence, a merciful creed, to be held by men endowed with pity and kindly feelings. It is therefore both unjust and deplorable that the word eugenics should be connected with Nazi racialist practices' (10). He advances similar arguments in *Eugenics: Galton and After.*
23. Naomi Mitchison, *Solution Three* (New York: Feminist Press at the City University of New York, 1995), p. 50.
24. Watson and Crick argued that DNA coded for RNA, which produced protein in a one-way sequence that could not be reversed. Crick described this as the 'central dogma' of genetics, although it subsequently turned out to be false. Crick describes the trouble he got into over his use of the word 'dogma' in his autobiography, *What Mad Pursuit: A Personal View of Scientific Discovery* (New York: Basic Books, 1988), p. 109.
25. Evelyn Fox Keller, *The Mirage of a Space between Nature and Nurture* (Durham, NC, and London: Duke University Press, 2010), p. 51.
26. Waddington develops this concept in detail in *The Strategy of the Gene: A Discussion of Some Aspects of Theoretical Biology* (London: Allen and Unwin, 1957).
27. Quoted in Susan Squier, *Babies in Bottles: Twentieth-Century Visions of Reproductive Technology* (New Brunswick, NJ: Rutgers University Press, 1994), p. 192.
28. For an account of this, see Connelly, p. 88.
29. Mark Singleton, 'Yoga, Eugenics and Spiritual Darwinism in the Early Twentieth Century', *International Journal of Hindu Studies* 11 (2007), 426.
30. As Singleton suggests, these narratives tapped into the belief that Aryan caste Hindus were the descendants of an Indo-European race and that their evolutionary 'awakening' held the key to the regeneration of the Indian nation.
31. Sarah Hodges, *Contraception, Colonialism and Commerce: Birth Control in South India, 1020–1940* (Aldershot: Ashgate, 2008), p. 124.

32. See Connelly, pp. 141–2, for a discussion of this.
33. Nehru, pp. 2–3.
34. Sanger, pp. 54–5.
35. Shri K. M. Munshi, 'Message of Good Wishes', in *The Third International Conference*, p. 4.
36. Quoted in Connelly, p. 229.
37. Quoted in Connelly, p. 318.
38. Quoted in Connelly, p. 321.
39. Damian Grant, *Salman Rushdie* (Tavistock: Northcote House, 1999), p. 52; Salman Rushdie, *Imaginary Homelands: Essays and Criticism 1981–1991* (London: Penguin, 1991), p. 16.
40. For an excellent overview, see Stephen Morton, *Salman Rushdie: Fictions of Postcolonial Modernity* (Basingstoke: Palgrave Macmillan, 2007); see also *The Cambridge Companion to Salman Rushdie*, ed. by Abdulrazak Gurhah (Cambridge: Cambridge University Press, 2007).
41. Clare Barker, 'Interdisciplinary Dialogues: Disability and Postcolonial Studies', *Journal of Literary and Cultural Disability Studies* 4 (2010), 219–36.
42. Salman Rushdie, *Midnight's Children* (London: Vintage, 2006), pp. 275–7.
43. Jacques Derrida, 'Telepathy', *Oxford Literary Review* 10 (1988), 3–41.
44. Royal Commission on Population *Report*, pp. 134–5.
45. See Paul, chapter 1.
46. Quoted in Paul, pp. 40, 55.
47. See, for example, Sir Charles Arden-Close, 'Our Crowded Island', *Eugenics Review* 40 (1948), 23–30.
48. Dudley Barker, *People for the Commonwealth: The Case for Mass Migration* (London: T. Wemer Laurie Ltd, 1949), pp. 5–6.
49. Alison Bashford, *Imperial Hygiene: A Critical History of Colonialism, Nationalism and Public Health* (Basingstoke: Palgrave Macmillan, 2003).
50. Laksiri Jayasuriya, 'From White Australia to Fortress Australia' in *Legacies of White Australia*, ed. by Laksiri Jayasuriya, David Walker and Jan Gothard (Crowley, WA: University of Western Australia Press, 2003), p. 58.
51. Quoted in Gwenda Tavan, *The Long, Slow Death of White Australia* (Carlton North, Victoria: Scribe Publications, 2005), pp. 34–5.
52. Stephen Garton, 'Eugenics in Australia and New Zealand: Laboratories of Racial Science', *The Oxford Handbook of the History of Eugenics*, ed. by Alison Bashford and Philippa Levine (New York: Oxford University Press, 2010), p. 248.
53. A. James Hammerton and Alistair Thomson, *Ten Pound Poms: Australia's Invisible Migrants* (Manchester: Manchester University Press, 2005), p. 266.
54. Elizabeth Jolley, *Central Mischief* (Ringwood, Victoria: Viking, 1992).
55. Quoted in *Ten Pound Poms*, p. 59.
56. Elizabeth Jolley, *The Vera Wright Trilogy* (New York: Persea Books, 2010), p. 494.

NOTES TO AFTERWORD

1. See C. P. Blacker, *Eugenics: Galton and After* (London: Duckworth, 1952), p. 13. For 'structures of feeling', see Raymond Williams, *The Long Revolution* (London: Hogarth Press, 1992), pp. 48–9.
2. Michel Foucault, *The Birth of Biopolitics: Lectures at the Collège de France 1978–1979* (Basingstoke: Palgrave Macmillan, 2008).
3. Hilary Rose, 'Eugenics and Genetics: The Conjoint Twins?', *New Formations* 60 (Spring 2007), 18.

4. Nikolas Rose, *The Politics of Life Itself: Biomedicine, Power, and Subjectivity in the Twenty-First Century* (Princeton, NJ: Princeton University Press, 2007), p. 63.
5. Nicholas Agar, *Liberal Eugenics: In Defence of Human Enhancement* (Oxford: Blackwell, 2004), p. 6.
6. Allen Buchanan, Dan W. Brock, Norman Daniels and Daniel Wikler, *From Chance to Choice: Genetics and Justice* (Cambridge: Cambridge University Press, 2000), p. 42.
7. Jürgen Habermas, *The Future of Human Nature* (Cambridge and Malden, MA: Polity Press, 2003), p. 13.
8. See, for example, Lenny Moss, 'Contra Habermas and towards a Critical Theory of Human Nature and the Question of Genetic Enhancement', *New Formations* 60 (Spring 2007), 139–49.
9. Michael Sandel, *The Case against Perfection: Ethics in the Age of Genetic Engineering* (Cambridge, MA, and London: Harvard University Press, 2007), pp. 82–3.
10. Troy Duster, *Backdoor to Eugenics*, 2nd edition (New York and London: Routledge, 2003), p. x.
11. These figures are taken from 'Down's Syndrome Trend Examined', available at <http://www.nhs.uk/news/2009/10October/Pages/Down-syndrome-termination-rate-screening.aspx> [accessed 18 January 2012] [paras. 7–10 of 20].
12. For a discussion of these issues, see Bill Armer, 'Eugenetics: A Polemical View of Social Policy in the Genetic Age', *New Formations* 60 (Spring 2007), 89–101.
13. Eva Hoffman, *The Secret* (London: Vintage, 2003), p. 225 (ellipsis in the original).
14. Sarah Franklin, *Dolly Mixtures: The Remaking of Genealogy* (Durham, NC, and London: Duke University Press), p. 14.
15. Matthew Eatough, 'The Time that Remains: Organ Donation, Temporal Duration, and *Bildung* in Kazuo Ishiguro's *Never Let me Go*', *Literature and Medicine* 29 (Spring 2011), 132–60.
16. Kazuo Ishiguro, *Never Let Me Go* (London: Faber and Faber, 2006), p. 156.
17. See Zygmunt Bauman, *Living on Borrowed Time: Conversations with Citlali Rovirosa-Madrazo* (Cambridge and Malden, MA: Polity Press, 2010), pp. 141–2.
18. Margaret Atwood, *Oryx and Crake* (London: Virago, 2004), p. 231.
19. Margaret Atwood, '*The Handmaid's Tale* and *Oryx and Crake* in Context', *PMLA* 119 (May 2004), 513–7.
20. David Macey, 'Rethinking Biopolitics, Race and Power in the Wake of Foucault, *Theory, Cutlure and Society* 26 (2009), 186–205.

Selected Bibliography

UNPUBLISHED SOURCES

C. P. Blacker Papers, Wellcome Library
Francis Crick Papers, Wellcome Library
Family Planning Association Papers, Wellcome Library
International Planned Parenthood Federation Papers, Wellcome Library
Patients Association Papers, Wellcome Library
The Norwood Committee Papers, The National Archives
C. H. Waddington Papers, Edinburgh University Library

PUBLISHED SOURCES

Agar, Nicholas, *Liberal Eugenics: In Defence of Human Enhancement* (Oxford and Victoria: Blackwell, 2004).
Alexander, Peter, *Atomic Radiation and Life* (Harmondsworth: Penguin, 1957).
Allingham, Margery, *The China Governess* (1963) (London: Penguin, 1965).
———. *The Mind Readers* (1965) (London: Penguin, 1968).
Arden-Close, Sir Charles, 'Our Crowded Island', *Eugenics Review* 40 (1948), 23–30.
Armer, Bill, 'Eugenetics: A Polemical View of Social Policy in the Genetic Age', *New Formations* 60 (2007), 89–101.
Atwood, Margaret, '*The Handmaid's Tale* and *Oryx and Crake* in Context', *PMLA* 119 (May 2004) 513–7.
Atwood, Margaret, *Oryx and Crake* (London: Virago, 2004).
Ballard, J. G., *The Drought* (1965) (London: Flamingo, 2002).
———. *Miracles of Life: Shanghai to Shepperton* (London: Fourth Estate, 2008).
Banton, Michael, *White and Coloured: The Behaviour of British People towards Coloured Immigrants* (London: Jonathan Cape, 1959).
Barker, Clare, 'Interdisciplinary Dialogues: Disability and Postcolonial Studies', *Journal of Literary and Cultural Disability Studies* 4 (2010), 219–36.
Barker, Dudley, *People for the Commonwealth: The Case for Mass Migration* (London: T. Wemer Laurie Ltd, 1948).
Bashford, Alison, *Imperial Hygiene: A Critical History of Colonialism, Nationalism and Public Health* (Basingstoke: Palgrave Macmillan, 2003).
Bashford, Alison and Philippa Levine, *The Oxford Handbook of the History of Eugenics* (New York: Oxford University Press, 2010).
Bauman, Zygmunt, *Living on Borrowed Time: Conversations with Citlalj Rovirosa-Madrazo* (Cambridge and Malden, MA: Polity Press, 2010).

Bertram, Colin, *Adam's Brood: Hopes and Fears of a Biologist*, Foreword by Sir Julian Huxley, F. R. S. (London: Peter Davies, 1959).

Bertram, G. C. L., *West Indian Immigration* (London: The Eugenics Society, 1958).

Beveridge, William, 'Eugenic Aspects of Children's Allowances' in *Twelve Galton Lectures: A Centenary Selection with Commentaries*, ed. by Steve Jones and Milo Keynes (London: Galton Institute, 2007).

Bhabha, Homi, 'Signs Taken for Wonders' in *The Location of Culture* (London and New York: Routledge, 1994).

Blacker, C. P., *Neurosis and the Mental Health Services* (1946) (London, New York, Toronto: Oxford University Press, 1948.

Blacker, C.P., '"Eugenic" Experiments Conducted by the Nazis on Human Subjects', *Eugenics Review* 44 (April 1952), 9–19.

Blacker, C. P. *Eugenics: Galton and After* (London: George Duckworth & Co. Ltd, 1952).

Blacker, C. P. (ed.), *Problem Families: Five Inquiries* (London: Eugenics Society, 1952).

Blacker, C. P., 'Voluntary Sterilization: Transitions throughout the World', *Eugenics Review* 54 (October 1962), 143–62.

Bland, Lucy, 'British Eugenics and "Race Crossing": A Study of an Interwar Investigation', *New Formations* 60 (2007), 66–78.

Bowlby, John, *Childcare and the Growth of Love* (Harmondsworth: Penguin, 1953)

Braine, John, *Room at the Top* (1957) (London: Arrow Books, 1997).

Brattain, Michelle, 'Race, Racism and Antiracism: UNESCO and the Politics of Presenting Science to the Postwar Public', *The American Historical Review* 112 (2007), available at <http://www.historycooperative.org/journals/ahr/112.5/brattain.html> [accessed 15 June 2011].

Bryan, Beverley, Stella Dadzie and Suzanne Scafe, *The Heart of the Race: Black Women's Lives in Britain* (London: Virago, 1985).

Buchanan, Allen, Dan W. Brock, Norman Daniels and Daniel Wikler, *From Chance to Choice: Genetics and Justice* (Cambridge: Cambridge University Press, 2000).

Burgess, Anthony, *A Clockwork Orange* (1961) (London: Penguin Classics, 2000).

———. *The Wanting Seed* (1962) (New York: W.W. Norton and Company, 1996).

Burt, C. L., 'The Evidence for the Concept of Intelligence', *British Journal of Educational Psychology* 15 (1955), 158–77.

———. 'The Inheritance of Mental Abilities', *American Psychology* 13 (1958), 1–15.

———. 'The Genetic Determination of Intelligence: A Study of Monozygotic Twins Reared Together and Apart', *British Journal of Psychology* 57 (1966), 137–53.

Carby, Hazel V., '"Schooling in Babylon" in the Centre for Contemporary Cultural Studies' in *The Empire Strikes Back: Race and Racism in 70s Britain* (Abingdon and New York: Routledge, 1994).

Carter, C. O., 'Promising Families: Some Conclusions', *Eugenics Review* 52 (1961), 197–200.

Cattell, Raymond B., *The Fight for Our National Intelligence*, with introductions by Lord Horder, Major Darwin and F. P. Armitage (London: P.S. King & Son, 1937).

Centre for Contemporary Cultural Studies, *The Empire Strikes Back: Race and Racism in 70s Britain* (Abingdon and New York: Routledge, 1994).

Chodorow, Nancy, *The Reproduction of Mothering: Psychoanalysis and the Sociology of Gender* (Berkeley: University of California Press, 1979).

Cockin, Katherine, and Jago Morrison (eds.), *The Post-war British Literature Handbook* (London: Continuum, 2009).
Collini, Stefan, 'Introduction' to C. P. Snow, *The Two Cultures* (Cambridge: Cambridge University Press, 1993).
Connelly, Matthew, *Fatal Misconception: The Struggle to Control World Population* (Cambridge, MA, and London: Harvard University Press, 2008).
Connor, Steven, *The English Novel in History: 1950 to the Present* (London and New York: Routledge, 1995).
Cooper, Richard S., Jay S. Kaufman and Ryk Ward, 'Race and Genomics', *New England Journal of Medicine* 348 (20 March 2003), 1166–70.
Crick, Francis, 'Discussion: Eugenics and Genetics' in *Man and His Future: A Ciba Foundation Volume*, ed. by Gordon Wolstenholme (Boston: Little, Brown & Co., 1963).
———. *What Mad Pursuit: A Personal View of Scientific Discovery* (New York: Basic Books, 1988).
Damian Grant, *Salman Rushdie* (Tavistock: Northcote House, 1999).
Darlington, C. D., 'The Genetic Understanding of Race in Man', *International Social Science Bulletin* 2 (Winter 1950), 487.
———. *The Facts of Life* (London: George Allen & Unwin, 1953).
———. *Genetics and Man* (Harmondsworth: Penguin, 1964).
Davis, John, *Youth and the Condition of Britain: Images of Adolescent Conflict* (London: Athlone Press, 1990).
Davis, Kingsley, 'Population Policy: Will Current Programs Succeed?', *Science* 158 (10 November 1967), 730–9.
Delaney, Shelagh, *A Taste of Honey* (London: Methuen, 1963).
Dench, Geoff (ed.), *The Rise and Rise of Meritocracy* (Oxford: Blackwell, 2006).
Depo-Provera: A Report (London: Campaign against Depo-Provera, 1983).
Derrida, Jacques, 'Telepathy', *Oxford Literary Review* 10 (1988), 3–41.
Douglas, Mary, *Purity and Danger* (1966) (London: Routledge Classics Edition, 2002).
Drabble, Margaret, *Jerusalem the Golden* (1967) (Harmondsworth: Penguin, 1972).
Duster, Troy, *Backdoor to Eugenics*, 2nd edition (New York and London: Routledge 2003).
Eathough, Matthew, 'The Time that Remains: Organ Donation, Temporal Duration, and *Bildung* in Kazuo Ishiguro's *Never Let me Go*', *Literature and Medicine* 29 (Spring 2011), 132–60.
Ehrlich, Paul, *The Population Bomb* (Cutchogue, NY: Buccaneer Books, 1971).
Emecheta, Buchi, *Second-class-Citizen* (1974) (London: Heinemann, 1994).
———. *In the Ditch* (London: Flamingo, 1988).
Eugenics Society Archive, Wellcome Library (SA/EUG/C/186).
Eysenck, H. J., *Uses and Abuses of Psychology* (Harmondsworth: Penguin, 1953).
———. *The IQ Argument: Race, Intelligence and Education* (New York: The Library Press, 1971).
Fanon, Franz, *Black Skin, White Masks*, trans. Charles Lam Markmann (London: Pluto Press, 1986).
Foucault, Michel, *Society Must Be Defended: Lectures at the Collège de France 1975–76*, trans. David Macey (London: Penguin, 2004).
———. *The Birth of Biopolitics: Lectures at the Collège de France, 1978–1979* (Basingstoke: Palgrave Macmillan, 2008).
———.'Truth and Power' in *The Foucault Reader*, ed. by Paul Rabinow (Harmondsworth: Penguin, 1986).
Fox Keller, Evelyn, *The Century of the Gene* (Cambridge, MA, and London: Harvard University Press, 2000).

Fox Keller, Evelyn, *Making Sense of Life: Explaining Biological Development with Models, Metaphors and Machines* (Cambridge, MA, and London: Harvard University Press, 2002).
———. *The Mirage of a Space between Nature and Nurture* (Durham, NC: Duke University Press, 2010).
Franklin, Sarah, *Dolly Mixtures: The Remaking of Genealogy* (Durham, NC, and London: Duke University Press, 2007).
Friedan, Betty, *The Feminine Mystique* (Harmondsworth: Penguin, 1963).
Fukuyama, Francis, *Our Posthuman Future: Consequences of the Biotechnology Revolution* (New York: Farrar, Straus & Giroux, 2002).
Galton, Francis, *Inquiries into Human Faculty and Its Development* (London: Macmillan, 1883).
———. *Essays in Eugenics* (London: Eugenics Education Society, 1909).
Garton, Stephen, 'Eugenics in Australia and New Zealand: Laboratories of Racial Science' in *The Oxford Handbook of the History of Eugenics*, ed. by Alison Bashford and Philippa Levine (New York: Oxford University Press, 2010).
Gasiorek, Andrzej. *Post-war British Fiction: Realism and After* (London: Hodder, 1995).
Gavron, Hannah, *The Captive Wife: Conflicts of Housebound Mothers* (London: Routledge and Kegan Paul Ltd, 1966).
Gissis, Snait B., and Eva Jablonka (eds), *Transformations of Lamarckism* (Cambridge, MA: MIT Press, 2011).
Gluckman, Peter, and Mark Hanson, *Mismatch* (Oxford: Oxford University Press, 2006).
Goffman, Erving, *Asylums: Essays on the Social Situation of Mental Patients and Other Inmates* (Penguin Social Studies) (London: Penguin, 1991).
Goldthorpe, John H., with Catriona Llewellyn and Clive Payne, *Social Mobility and Class Structure in Modern Britain*, 2nd edition (Oxford: Clarendon Press, 1980).
Gorer, Geoffrey, 'The Perils of Hypergamy' in *The Danger of Equality and Other Essays* (London: Cresset Press, 1966).
Greenhalgh, Susan, 'The Social Construction of Population Science: An Intellectual, Institutional, and Political History of Twentieth-Century Demography', *Comparative Studies in Society and History* 38 (January 1996), 26–66.
Gurhah, Abdulrazak (ed.), *The Cambridge Companion to Salman Rushdie* (Cambridge: Cambridge University Press, 2007).
Habermas, Jürgen, *The Future of Human Nature* (Cambridge and Malden, MA: Polity Press, 2003).
Hall, Lesley A., 'Women, Feminism and Eugenics' in *Essays in the History of Eugenics*, ed. by Robert A. Peel (London: Galton Institute, 1998).
Hammerton, A. James, and Alistair Thomson, *Ten Pound Poms: Australia's Invisible Migrants* (Manchester: Manchester University Press, 2005).
Hanson, Clare, *A Cultural History of Pregnancy: Pregnancy, Medicine and Culture, 1750–2000* (Basingstoke and New York: Palgrave Macmillan, 2004).
Harman, Oren Solomon, *The Man Who Invented the Chromosome: A Life of Cyril Darlington* (Cambridge, MA, and London: Harvard University Press, 2004).
Hearnshaw, L. S., *Cyril Burt: Psychologist* (Ithaca, NY: Cornell University Press, 1979).
Heath, Anthony, and Clive Payne, 'Twentieth Century Trends in Social Mobility in Britain', Centre for Research into Elections and Social Trends, Working Paper No. 70, June 1999.
Hilliard, L. T., 'Resettling Mental Defectives: Psychological and Social Aspects', *British Medical Journal* 1 (1954), 1372–4.

Hilliard, L. T., and Brian Kirman, *Mental Deficiency*, 2nd edition (London: J. & A. Churchill Ltd., 1965).
Hodges, Sarah, *Contraception, Colonialism and Commerce: Birth Control in South India, 1020–1940* (Aldershot: Ashgate, 2008).
Hoffman, Eva, *The Secret* (London: Vintage, 2003).
Hoggart, Richard, *The Uses of Literacy: Aspects of Working-Class Life with Special Reference to Publications and Entertainments* (1957) (Harmondsworth: Penguin, 1962).
Hubback, Eva M., *The Population of Britain* (Harmondsworth: Penguin, 1947).
Hubback, Judith, *Wives Who Went to College* (London: Heinemann, 1957).
Husband, Charles (ed.), *'Race' in Britain: Continuity and Change* (London: Hutchinson, 1982).
Huxley, Aldous, *Brave New World* (1962) (London: Flamingo, 1994).
———. 'The Double Crisis', *The Science News-Letter* 55 (1932) (26 March 1949), 199–202.
———. *Brave New World Revisited* (1959) (London: Vintage, 2004).
———. *Island* (London: Vintage, 2004).
Huxley, Julian. *Evolution: The Modern Synthesis* (London: George Allen and Unwin, 1942).
———. 'Soviet Genetics: The Real Issue', *Nature* 163 (June 1949), 935–42.
———. 'World Population', *Scientific American* 194 (1956), 66–9.
———. *New Bottles for New Wine* (London: Chatto & Windus, 1957).
———. *Essays of a Humanist* (Harmondsworth: Penguin, 1966).
Huxley, Julian (ed.), *The Humanist Frame* (London: Allen and Unwin, 1961).
International Planned Parenthood Federation home page, available at <http://www.ippf.org/en/> [accessed 15 June 2011].
IPPF papers, Wellcome Library, PP/MCS/B.9.
Ishiguro, Kazuo, *Never Let Me Go* (London: Faber and Faber, 2006).
Ittmann, Karl, 'Demography as Policy Science in the British Empire, 1918–1969', *Journal of Policy History* 15 (2003), 417–48.
Jayasuriya, Laksiri, 'From White Australia to Fortress Australia' in *Legacies of White Australia*, ed. by Laksiri Jayasuriya, David Walker and Jan Gothard (Crowley, WA: University of Western Australia Press, 2003).
Jensen, Arthur S., 'How Much Can We Boost IQ and Scholastic Achievement?', *Harvard Educational Review* 33 (1969), 1–123.
Johnson, Kelley, and Rannveig Traustadóttir (eds), *Deinstitutionalisation and People with Intellectual Disabilities: In and Out of Institutions* (London and Philadelphia: Jessica Kingsley Publishers, 2005).
Jolley, Elizabeth, *My Father's Moon* (New York: Harper & Row, 1989).
———. *Cabin Fever* (New York: Harper & Row, 1990)
———. *Central Mischief* (Ringwood, Victoria: Viking, 1992).
———. *The Georges' Wife* (Camberwell, Victoria: Penguin Books Australia, 1993).
———. *The Vera Wright Trilogy* (New York: Persea Books, 2010).
Jones, Greta, *Social Darwinism and English Thought: The Interaction between Biological and Social Theory* (Brighton: Harvester Press, 1980).
Jones, Kathleen, *A History of the Mental Health Services* (London and Boston: Routledge and Kegan Paul, 1972).
Karabel, J., and A. H. Halsey (eds), *Power and Ideology in Education* (Oxford and New York: Oxford University Press, 1977).
Kay, Lily E., *Who Wrote the Book of Life: A History of the Genetic Code* (Stanford, CA: Stanford University Press, 2000).
Kevles, Daniel J., *In the Name of Eugenics: Genetics and the Uses of Human Heredity*, with a new preface by the author (Cambridge, MA, and London: Harvard University Press, 1996).

———. 'Eugenics and Human Rights', *British Medical Journal* 319 (14 August 1999), 435–8.

Kristeva, Julia, 'Approaching Abjection' in *Powers of Horror, The Portable Kristeva*, ed. by Kelly Oliver (Chichester and New York: Columbia University Press, 1997).

Laing, R. D., *The Divided Self: An Existential Study in Sanity and Madness* (London: Penguin, 1990).

Lamb, Marion J., 'Attitudes to Soft Inheritance in Great Britain, 1930s–1970s', in *Transformations of Lamarckism*, ed. by Snait B. Gissis and Eva Jablonka (Cambridge, MA: MIT Press, 2011).

Lamming, George, *The Emigrants* (1954) (London and New York: Allison and Busby, 1980).

Land, Hilary, 'We Sat Down at the Table of Privilege and Complained about the Food' in *The Rise and Rise of Meritocracy*, ed. by Geoff Dench (Oxford: Blackwell, 2006), pp. 45–60.

Laura Otis (ed.), *Literature and Science in the Nineteenth Century: An Anthology* (Oxford: Oxford World's Classics, 2002).

Lawrence, Errol, 'In The Abundance of Water the Fool is Thirsty: Sociology and Black "Pathology"', in Centre for Contemporary Cultural Studies, *The Empire Strikes Back: Race and Racism in 70s Britain* (Abingdon and New York: Routledge, 1994).

Lecourt, Dominique, *Proletarian Science?: The Case of Lysenko*, with an introduction by Louis Althusser, trans. Ben Brewster (London: New Left Books, 1977).

Lessing, Doris, *The Four-Gated City: Volume Five of Children of Violence* (London: MacGibbon & Kee, 1969).

———. 'In the World, Not of It', *Encounter* (August 1972), 61–4.

———. 'If You Knew Sufi', *Guardian* (8 January 1975).

———. *Re: Colonised Planet 5 Shikasta* (1979) (London and New York: Granada, 1981).

———. *The Sirian Experiments* (1980) (London: Flamingo, 1994).

———. 'Learning to Put the Questions Differently' in *Putting the Questions Differently: Interviews with Doris Lessing*, ed. by Earl G. Ingersoll (London: Flamingo, 1996).

———. *Walking in the Shade: Volume Two of My Autobiography, 1949–1962* (London: Flamingo, 1998).

Lewis, Aubrey, 'Psychopathic Personality: A Most Elusive Category', *Psychological Medicine* 4 (1974), 133–40.

Light, Alison, *Forever England: Femininity, Literature and Conservatism between the Wars* (London and New York: Routledge, 1991).

Luckhurst, Roger, *'The Angle between Two Walls': The Fiction of J.G. Ballard* (Liverpool: Liverpool University Press, 1997).

Mace, D. R., *Marriage Counselling: The First Full Account of the Remedial Work of the Marriage Guidance Counsellors* (London: J. & A. Churchill Ltd, 1948).

Macey, David, 'Rethinking Biopolitics, Race and Power in the Wake of Foucault', *Theory, Culture and Society* 26 (2009), 186–205.

MacInnes, Colin, *Colin MacInnes Omnibus* (London: Allison and Busby, 2005).

Marks, John, 'The New Eugenics: Jacques Testart and French Bioethics', *New Formations* 60 (2007), 124–38.

Marshall, T. H., 'Social Selection in the Welfare State', *Eugenics Review* 45 (1953), 81–92.

Mayr, Ernst, 'Change in Genetic Environment and Evolution' in *Evolution as a Process*, ed. by Julian Huxley, A. C. Hardy and E. B. Ford (London: George Allen and Unwin, 1954), pp. 157–80.

Mazumdar, Pauline, *Eugenics, Human Genetics and Human Failings: The Eugenics Society, Its Sources and Its Critics in Britain* (London: Routledge, 1992).

McLaren, Anne, 'Dealing with Life' in *What the Human Race Is Up To*, ed. by Naomi Mitchison (London: Gollancz, 1962).

Medawar, Peter, *The Future of Man* (New York: Mentor Books, 1961).

Mitchison, Naomi, *Memoirs of a Spacewoman* (1962) (London: New English Library, 1976).

———. *All Change Here: Girlhood and Marriage* (London: Bodley Head, 1975).

———. *Solution Three* (New York: Feminist Press at the City University of New York, 1995).

Mitchison, Naomi (ed.), *What the Human Race Is Up To* (London: Gollancz, 1962).

Morris, Pauline, *Put Away: A Sociological Study of Institutions for the Mentally Retarded*, with a Foreword by Peter Townsend (London: Routledge and Kegan Paul, 1969).

Mortimer, Penelope, *The Pumpkin Eater* (1962) (Harmondsworth: Penguin, 1964).

Morton, Stephen, *Salman Rushdie: Fictions of Postcolonial Modernity* (Basingstoke: Palgrave Macmillan, 2007).

Moss, Lenny, 'Contra Habermas and Towards a Critical Theory of Human Nature and the Question of Genetic Enhancement', *New Formations* 60 (2007), 139–49.

Muller, H. J., 'Comments and Criticisms on Different Items of the Statement' in *The Race Concept: Results of an Inquiry* (Paris: UNESCO, 1952), p. 52.

Myrdal, Alva, and Viola Klein, *Women's Two Roles: Home and Work* (1956), 2nd revised edition (London: Routledge and Kegan Paul Ltd, 1968).

Nichols, Peter, *A Day in the Death of Joe Egg* (London: Faber and Faber, 1967).

Osbergy, Bill, *Youth in Britain since 1945* (Making Contemporary Britain) (Oxford: Wiley-Blackwell, 1997).

Osborne, John, *Look Back in Anger* (1957) (London: Faber and Faber Ltd, 1983).

Patterson, Sheila, *Dark Strangers: A Study of West Indians in London* (Harmondsworth: Penguin, 1965).

Paul, Diane C., 'A War on Two Fronts: J. B. S. Haldane and the Response to Lysenkoism in Britain', *Journal of the History of Biology* 16 (Spring 1983), 1–37.

Paul, Kathleen, *Whitewashing Britain: Race and Citizenship in the Postwar Era* (Ithaca, NY, and London: Cornell University Press, 1997).

Powell, Enoch, 'Rivers of Blood' speech, reprinted in *The Daily Telegraph*, available at <http://www.telegraph.co.uk/comment/3643823/Enoch-Powells-Rivers-of-Blood-speech.html> [accessed 7 April 2011].

Powell, Enoch J., and Rex Collings (eds), *Reflections: The Selected Writings and Speeches of Enoch Powell* (London: Bellew Publishing Co. Ltd, 1992).

Priest, Christopher, *Fugue for a Darkening Island* (London: Faber and Faber, 1972).

The Race Concept: Results of an Inquiry (Paris: UNESCO, 1952), p. 52.

Report on the Effect of Radiation on Human Heredity (Geneva: World Health Organization, 1957).

Richardson, Angelique, *Love and Eugenics in the Late Nineteenth Century: Rational Reproduction and the New Woman* (Oxford: Oxford University Press, 2003).

Roll-Hansen, Nils, 'Lamarckism and Lysenkoism Revisited' in *Transformations of Lamarckism*, ed. by Snait B. Gissis and Eva Jablonka (Cambridge, MA: MIT Press, 2011).

Rose, Hilary, 'Eugenics and Genetics: The Conjoint Twins?', *New Formations* 60 (2007), 13–26.

Rose, Nikolas, 'The Politics of Life Itself', *Theory, Culture and Society* 18 (2001), 1–30.

———. *The Politics of Life Itself: Biomedicine, Power, and Subjectivity in the Twenty-First Century* (Princeton, NJ: Princeton University Press, 2007).

Rose, Steven, review of Arthur Jensen, *Educability and Group Differences, New Scientist* (14 June 1973), 704–5.

———. 'Hans Eysenck's Controversial Career', *The Lancet* 376 (7 August 2010), 407–8.

Rowbotham, Sheila, *Threads through Time: Writings on History and Autobiography* (Harmondsworth: Penguin, 1999).

Royal Commission on Population *Report* (London: HMSO, 1949).

Rushdie, Salman, *Imaginary Homelands: Essays and Criticism 1981–1991* (London: Penguin, 1991).

Rushdie, Salman, *Midnight's Children* (London: Vintage, 2006).

Ryan, Joanna, with Frank Thomas, *The Politics of Mental Handicap* (Harmondsworth: Penguin, 1980).

Sage, Lorna, *Bad Blood* (2000) (London: Fourth Estate, 2001).

Sandbrook, Dominic, *Never Had It So Good: A History of Britain from Suez to the Beatles* (London: Little, Brown, 2005).

Sandel, Michael, *The Case against Perfection: Ethics in the Age of Genetic Engineering* (Cambridge, MA, and London: Harvard University Press, 2007).

Schenk, Faith, and A. S. Parkes, 'The Activities of the Eugenics Society', *Eugenics Review* 60 (September 1968), 142–61.

Seed, David, *American Science Fiction: Literature and Film* (Edinburgh: Edinburgh University Press, 1999).

Selvon, Sam, *The Lonely Londoners* (1956) (London: Penguin Modern Classics, 2006).

Singleton, Mark, 'Yoga, Eugenics and Spiritual Darwinism in the Early Twentieth Century', *International Journal of Hindu Studies* 11 (2007), 125–46.

Smith, Anna Marie, *New Right Discourse on Race and Sexuality: Britain, 1968–1990* (Cambridge: Cambridge University Press, 1994).

Snow, C. P., *The New Men* (Harmondsworth: Penguin, 1959).

Soloway, Richard A., *Demography and Degeneration: Eugenics and the Declining Birthrate in Twentieth-Century Britain* (Chapel Hill and London: University of North Carolina Press, 1995).

Spark, Muriel, *The Ballad of Peckham Rye* (1960) (London: Penguin Modern Classics, 2006).

———. *The Go-Away Bird and Other Stories* (Harmondsworth: Penguin, 1963).

Squier, Susan Merrill, *Babies in Bottles: Twentieth Century Visions of Reproductive Technology* (New Brunswick, NJ: Rutgers University Press, 1994).

Steedman, Carolyn, *Landscape for a Good Woman: A Story of Two Lives* (London: Virago, 1986).

Strathern, Marilyn, *Reproducing the Future: Anthropology, Kinship and the New Reproductive Technologies* (Manchester: Manchester University Press, 1992).

Swanson, Gillian, 'Serenity, Self-regard and the Genetic Sequence: Social Psychiatry and Preventive Eugenics in Britain, 1930s–1950s', *New Formations* 60 (2007), 50–65.
Tavan, Gwenda, *The Long, Slow Death of White Australia* (Carlton North, Victoria: Scribe Publications, 2005).
Thane, Pat, *Foundations of the Welfare State*, 2nd edition (Harlow: Pearson Education, 1996).
Third International Conference on Planned Parenthood: Report of the Proceedings (Bombay: Family Planning Association of India, 1952).
Thomson, Mathew, *The Problem of Mental Deficiency: Eugenics, Democracy, and Social Policy in Britain c.1870–1959* (Oxford: Clarendon Press, 1998).
Titmuss, Richard, *Poverty and Population: A Factual Study of Contemporary Social Waste* (London: Macmillan & Co., 1938).
———. *Essays on 'The Welfare State'* (London: George Allen and Unwin Ltd, 1958).
Todes, Daniel, 'Global Darwin: Contempt for Competition', *Nature* 7269 (November 2009), 36–7.
Tredgold, A. F, *A Textbook of Mental Deficiency*, 10th edition, ed. by R. F. Tredgold and K. Soddy (London: Baillière, Tindall & Cox, 1963).
UNESCO and Its Programme, Vol. 3, The Race Question (Paris: UNESCO, 1950).
Waddington, C. H., 'Lysenko and the Scientists', *The New Statesman and Nation* (25 December 1948), 566.
———. *The Strategy of the Gene: A Discussion of Some Aspects of Theoretical Biology* (London: Allen and Unwin, 1957).
Warnock, Mary, 'Special Educational Needs: A New Look' (London: Philosophy of Education Society of Great Britain, 2005).
Warnock Report: Special Educational Needs (London: HMSO, 1978).
Waters, Chris, '"Dark Strangers" in Our Midst: Discourses of Race and Nation in Britain, 1947–1963', *The Journal of British Studies* 36 (April 1997), 207–38.
Watson, James D., *The Double Helix*, with a new introduction by Steve Jones (London: Penguin, 1999).
———. 'Genetics and Eugenics: A Personal Odyssey' in *Medicine after the Holocaust: From the Master Race to the Human Genome*, ed. by Sheldon Rubenfeld, in conjunction with the Holocaust Museum, Houston (Basingstoke: Palgrave, 2010).
Wheatley, Denis, *To the Devil—a Daughter* (London: Hutchinson, 1953).
Wilkins, Leslie, *Delinquent Generations: A Home Office Research Unit Report* (London: HMSO, 1960)
Williams, Raymond, *Culture and Society 1780–1950* (1958) (Harmondsworth: Penguin, 1962).
———. *Border Country* (1960) (Cardigan: Parthian Library of Wales, 2006).
———. *The Long Revolution* (1961) (London: Hogarth Press, 1992).
Winnicott, D. W., *The Child, the Family and the Outside World* (Harmondsworth: Penguin, 1964).
Woodruffe, Sir John, *The Seed of Race: An Essay on Indian Education* (Madras: Ganesh & Co., 1919).
Wyndham, John, *The Day of the Triffids* (1951) (London: Penguin Classics, 1999).
———. *The Chrysalids* (1955) (London: Penguin, 1958).
———. *The Midwich Cuckoos* (1957) (London: Penguin, 2008).

Yallop, David, *To Encourage the Others* (London: W. H. Allen, 1971).
Young, Michael, *The Rise of the Meritocracy: An Essay on Education and Equality* (1958) (Harmondsworth: Penguin, 1961).
———. 'Down with Meritocracy', *The Guardian* (29 June 2001), available at http://www.guardian.co.uk/comment/story/> [accessed 2 August 2007].
———. 'Looking Back on *Meritocracy*', in Dench, pp. 73–7.
Yudkin, S., and A. Holme, *Working Mothers and Their Children* (London: Michael Joseph, 1963).

Index